CW01669681

For Rhona (xxx)

Sir Arthur Munro Sutherland

The Legend of Newcastle

(a good Sutherland)

Nigel McMurray

Nigel.

ISBN 9781659997972

Layout by Techstar, Jedburgh, Scottish Borders. www.jedpcrepair.co.uk

Cover design by Tara Maybury of Teabury Creative, Idaho, USA. teaburycreative.com

This book was funded by a legacy from the late Frances C. Truman of Waterloo, Iowa, USA.

Dedicated to the memory of my mother, Vera, a true Geordie.

Newcastle Mansion House Plaque. Photograph by Nigel McMurray.

Front cover image. *Sir Arthur Munro Sutherland* by Frank O. Salisbury.
By permission of the Estate of Frank O. Salisbury.

Pre-launch photograph of the "Sutherland" at Doxford's of Sunderland in 1934. This heralded the end of the Great Depression in the North East. Under licence from Tyne & Wear Archives.

Contents

Author's Notes.

What is in a name? Sutherland can refer to a Clan, the ancient Scots County or a surname. There are six identifiable generations of Sutherlands relating to Sir Arthur Sutherland and the Christian name Benjamin appears in four of those. Benjamin John (Sir Arthur's father) named his company BJ Sutherland which adds to the confusion as to whether one is referring to the individual or the company. Sir Arthur Munro Sutherland had a son of the same name with the family circumventing this problem by referring to the father as Arthur and the son as Munro, but an element of confusion remains relating to their different business interests. Sir Arthur was knighted in 1920 and so I have referred to him as Arthur prior to that date and Sir Arthur subsequent to it but this has not always been possible. He named seven ships "Sutherland" which could have referred to the ancient County or himself; their being named after the County is viable given that other ships were named Caithness or Dunrobin, places of significance to the area. Some of his philanthropic enterprises were named Sutherland after other members of the family. Every effort has been made in the text to differentiate the various references to Sutherland, but elements of confusion remain.

Church or chapel? John Wesley referred to his early centres of worship as "rooms" or "houses" and not as "chapels" or "churches". The term "chapel" as a means of differentiating Nonconformist place of worship from the Established Church of England appeared after his death. Very broadly speaking, Methodist places of worship became known as "churches" from about 1932 (the date of the union of most of the Methodist denominations) but this is not a hard and fast rule. Four Methodist places of worship feature significantly in this monograph. The term Brunswick Methodist Church in Newcastle was first used outside that building in the late 1960s though Methodist Chapel is cut in stone on its front pediment. Sir Arthur was a lifelong member of the Clayton Road Methodist Chapel which was also known as Clayton Road Wesleyan Chapel (until Methodist Union in 1932) and as a Clayton Road Wesleyan *Church* in a commercial postcard published before 1932! The first Methodist building for worship was the New Room in Bristol by which name it is still called. Sir Arthur married for the second time (subsequent to his becoming a widower) in Wesley's Chapel in London by which name it is still known albeit with "A Methodist Church" being printed above its outside noticeboards. Consequently, within the text, different names may be given for the same building.

Beginnings

Sir Arthur Munro Sutherland (1867-1953) was a Newcastle upon Tyne shipowner, ship-broker, coal exporter and philanthropist known as "The Legend of Newcastle". The Honorary Freedom of the City of Newcastle was conferred upon him in 1936 in a lavish ceremony at its City Hall. The Freedom was conferred "in recognition of the outstanding public service that he has rendered to this City" and it was stated that "of the long role of those granted the Honorary Freeman of the City none has done more for Newcastle than Sir Arthur". His obituary in "The Novocastrian" (the magazine of the old boys association of the Royal Grammar School) stated; "There are few men who, in their own lifetime, are regarded as a legend, a reality, and at the same time a living embodiment of simplicity, shrewdness, and generosity".

Benjamin Sutherland (1799-1879)

Arthur's grandfather Benjamin Sutherland was born in Orkney but moved to Thurso on the Caithness mainland to carry out his trade as a boot-maker. When he was twenty-six, he left Thurso and sailed to Newcastle on one of the first steam ships to export Caithness flagstone. (These flagstones were then being exported to many of the developing cities in Scotland and northern England, including Newcastle for what is now known as Grainger Town.) The reason for his emigration is unknown but was probably related to the one hundred years of the Highland Clearances, particularly the Sutherland Clearances, with the subsequent mass emigration impacting on all aspects of life and trades in the Highlands. The harsh winters of those times saw many land workers head for the cities.

Quite why Benjamin chose Newcastle is unknown, but Newcastle was quite an attractive proposition as it was already a thriving town with many Scots immigrants, including members of the Haggie family whom he knew from Orkney. Benjamin's trade as a bootmaker was well suited to Tyneside in that his robust products for agricultural work in the north of Scotland were easily adapted to the requirements of the industrial workers in the shipyards and on the Quayside.

Once he was established in Newcastle he was joined by Alison Johnson (or Johnstone) Munro of Caithness and they married in the parish church of St John the Baptist in Grainger Street in 1825. (Solemnization by the Church of

England was required until the Marriage Act of 1836.) The next thirty years saw the family progress from near poverty to a lavish lifestyle as illustrated by the size and location of their homes. Although for years the family could only afford to live in the Northumberland House Star Hotel in the Byker St Peter's area of Newcastle, the family gradually started on the road to financial prosperity when they purchased their first home, a terrace house in Kent Street, Jarrow. They then moved to Newcastle's Northumberland Street and finally to Shieldfield Green at a time when Shieldfield was known as "a most pleasant and airy retreat for men of business". By 1858 he was able to employ thirteen men at his "shoe business in the centre of town" at 28 Northumberland Street. His company made bespoke footwear including riding boots, shoes including lawn tennis shoes, and imported high class products from London and Paris. It had been a long but ultimately successful journey from destitute Thurso.

Benjamin John Sutherland (1833-1901)

Alison and Benjamin had nine children the fourth being Benjamin John Sutherland who would prove to be an even more successful businessman than his father. He was educated at John Bruce's eminent Percy Street Academy and studied chemistry at the Rye Hill Colleges (now Newcastle College and not to be confused with the Royal Grammar School when it was at Rye Hill). He then worked as a lawyer's clerk, six years for his family friends the Haggie Brothers (by then rope-makers and timber importers in Gateshead) and then as a director of a building society.

Robert Verdin. "The King of Salt."

Photograph by Andrew Wright.

He decided to put all the diverse skills that he had acquired to good use by founding his own business in Broad Chare on the Quayside in 1857 as a chemical broker with links to the burgeoning chemical industry in Cheshire. The chemical industry was already thriving on Tyneside, but he chose not to become involved in it directly but to import products instead as a broker. The basis of his first success was fortuitous. A large cargo of sulphur had been wrecked off the Tyne in 1858 and was discarded as being useless. But making use of his education in chemistry at the Percy Street Academy and by public experiment, he showed that sulphur was undamaged by sea water. He then sold it at auction to the very men who had considered it of no value and so stole a march on his experienced business rivals. He soon began to import salt, sulphur ore and manganese from the two largest producers in Cheshire (Thomas Verdin & Sons of Northwich and Winsford – the "Kings of Salt" - and the Wheelock Iron & Salt Co. of Sandbach) for use in Tyneside chemical works. The chemical trade was where his first fortune was made as he made full use of the then cheap rail and sea links with areas outside the North East.

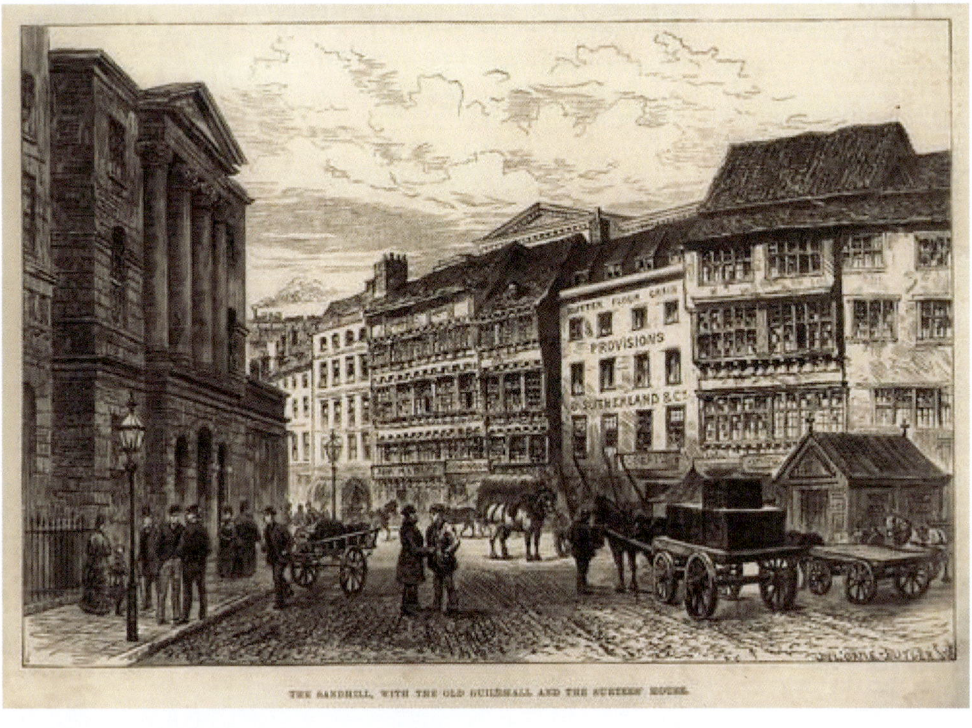

The Sandhill, Newcastle. C1880. By permission of Newcastle City Library.

The North East chemical industry went into gradual decline from the 1880s. Perceiving this, and being ever the entrepreneur, Benjamin John then branched out and became a butter, flour and grain merchant having purchased the "Butler's Flour, Grain & Provisions" company. He soon developed a virtual monopoly on the feed for the hundreds of horses used for haulage on the Quayside. The Corn Laws of 1790 prevented the importation of grain until local prices were prohibitively high. This protected the agricultural interest but not the growing industrial economy whose workers could barely afford to eat because of the high price of bread. The Corn Laws were repealed in 1846 allowing the importation of corn - grains that need grinding, especially wheat - primarily from North America without the payment of import duty. But it took another twenty-five years before improvements in farm machinery, railroads and shipping in the USA would enable the dumping of vast quantities of cheap corn on the European market. The Mid-Westerners struggled for years to prise open the doors of the European and British markets for their flour from their base in Minneapolis, Minnesota until finally Washburn-Crosby's Mills (later General Mills) of Minneapolis became "the milling centre of the world" and the "world's largest grain exporter".

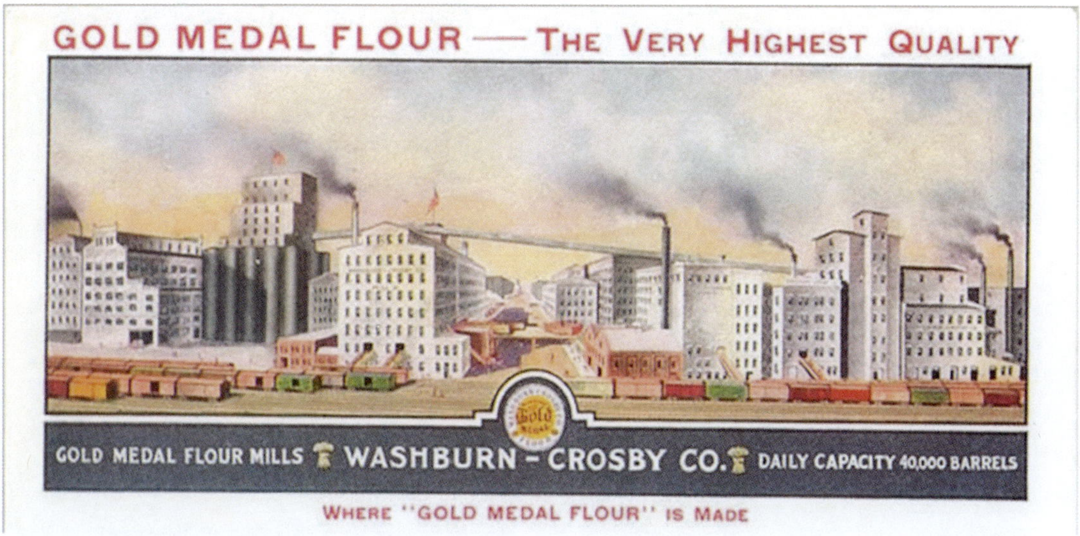

Washburn Crosby Flour Millers, Minneapolis. Washburn Crosby Company Postcard. c1900. Public Domain

William Hood Dunwoody, their Scots "foreign representative" to the UK in 1877, was given short shrift by British millers who would brook no "meddlesome interference" into their trade. They knew nothing of the new Minneapolis "gradual reduction" system of grinding the wheat between porcelain and iron rollers instead of abrasive millstones. And then using a "middlings purifier" to blast air through the flour to remove undesirable bran and husks. The product was lighter in colour and more visibly attractive to the extent that the UK rivals of Washburn-Crosby claimed that it must have been mixed with marble or dust to give the resultant flour its pale colour. Only a few hundred barrels were sold in 1877 and so on his next visit he approached the British bakers directly; the bakers confirmed that American flour was stronger and made more loaves per barrel than their native product. By 1895 over four million barrels were being exported to Britain from the endless supplies of wheat in the Mid-West. Benjamin John was one of the first British millers to see the important potential of this and he became a British Agent for Washburn-Crosby. In that capacity, he made eight visits to Canada and the Mid-West and became a major British importer of grain, flour and cheese from those countries into the UK. He also imported the first direct cargo of flour from San Francisco.

To cope with these imports he built his own mill, the Gateshead Steam Flour Mill adjacent to Hillgate Quay. All that remains of this mill is a copper plaque which is preserved in Gateshead Archives. Business was transacted in the Corn Exchange in Newcastle town hall. This, his second fortune, was made against the odds and the derision of fellow British flour millers.

Quilted Camel Retro Bar. 2020. Photograph by Nigel McMurray.

In 1857 Benjamin John moved his main office from Broad Chare on the bustling Quayside to a prestigious business property next door to Bessie Surtees House in the Sandhill. He was now opposite the Commercial Exchange in the Guildhall, known locally as The 'Change. His modern warehouses and administrative buildings were a few doors away in is now the Quilted Camel Retro Bar and the building behind it. Thus, by the 1880s his thriving business headquarters were at the heart of the Quayside and the outside wall proudly stated; "Flour, grain, provisions. BJ Sutherland & Co." In the 1891 Census Return he described himself as a "Flour and provisions merchant and miller".

Offices of Arthur Sutherland's shipping business alongside his father's original businesses.

By permission of the Sutherland family.

Finally, towards the end of his life and at the instigation of his son, Arthur, he moved somewhat tenuously into the shipping trade as shown by the initial inauspicious white raised lettering in one window of his offices; "BJ Sutherland Steamship Co. Ltd." Thus, the fortunes made in the chemical and grain trades were used to enable his son to make his own fortune in shipping. His final description of his business interests in "Tyneside Industries" was that of a "steamship owner and broker, flour and grain importer, coke and coal exporter". It appears that Benjamin John Sutherland, James Sutherland (a brother of Benjamin John) and Arthur were partners in the steamship company but after Benjamin John's sudden and unexpected death it was Arthur who took control.

Benjamin John became involved in local politics as a Liberal-Unionist and entered the Council as one of the Members for the St Nicholas Ward in 1887. He enjoyed a meteoric rise to fame being unanimously appointed Sheriff of Newcastle as early as 1891.

Benjamin John Sutherland as Sheriff of Newcastle in 1891. By permission of the Sutherland family.

As to faith issues, he had been baptised at the parish church of St Nicholas (now St Nicholas Cathedral). In 1859 he married Mary Ann Proud (1835-1901) the daughter of Joseph Proud, a grocer and Primitive Methodist "Ranter" preacher of Carrs Hill in Gateshead who played an integral part in local outdoor evangelical or 'camp meetings'. This was a highly unusual match of a rich Wesleyan capitalist, with a Primitive who aligned herself with the poor. It is the first *documented* evidence of the Sutherlands becoming actively involved with the Methodists and it would prove to be a life-long involvement. They were married in the Brunswick Place Wesleyan Methodist Chapel and would become life-long members of the Clayton Road Wesleyan Methodist Chapel in Jesmond from the date of its opening in 1883. (For a more detailed explanation of the branches of Methodism see **Methodism**.)

Initially, Benjamin John and family were also residents of Shieldfield Green but soon moved to Oxford Street in the centre of town. His businesses were thriving which enabled him to employ three servants - a nurse, cook and housemaid - plus a governess to look after the children. Ongoing success enabled him to purchase "Jesmond Vale" a standalone large dwelling with a wide, long drive then in Jesmond Vale Lane. He acquired an extensive property portfolio of five properties in the Sandhill (inherited by his sons) and six houses in Jesmond (inherited by his wife and daughters). The family's final home was Kelso House, a new semi-detached villa purchased in 1879 in Fernwood Road. (The name Kelso derived from that being the birthplace of James Archbold, the landowner from 1821, and the Archbold clan originating there from Norman times.) Archbold Street (now Terrace) was also named after him. The Sutherland family lived in one half of Kelso House which they renamed Thurso House reflecting their own ancestral origins. The other half retained the name Kelso House and was let until 1915 when Arthur became the sole owner, joined the two halves, and removed the name Kelso with which he had no connections. This is the building which would become Newcastle's Mansion House.

Arthur Munro Sutherland (1867-1953)

Mary and Benjamin John had ten children of whom the seventh, their fourth son, was Arthur Munro, Munro being his grandmother's maiden name. Arthur's father, as a rising star in the firmaments of business and politics, could well afford to send Arthur to the Royal Grammar School which he attended from 1878 to 1883 when it was based in Rye Hill (prior to its move to Jesmond in 1906).

Arthur Sutherland aged 18 months, with his nurse. By permission of the Northumberland Press Ltd.

But an academic career was never to Arthur's liking and he left at age sixteen. He later stated that he wished he'd left at fifteen. Two of Arthur's brothers also attended but for even shorter periods. Of the ten children it would be Arthur who would fully understand and grasp what his father and grandfather had achieved through perseverance, risk, overcoming opposition, financial acumen, diversification and developing national and international links for businesses.

THE SCHOOL AT RYE HILL, 1870-1906.

Royal Grammar School when located at Rye Hill. By permission of the Sutherland family.

Shipping Companies

His early departure from school at age sixteen was to his father's chagrin particularly as he ended up as an office boy with another Quayside firm, Gracie & Co. (Leith and Newcastle ship-brokers) "earning a few shillings a week and addressing envelopes and licking stamps". Nevertheless, he must have been more successful than he made out in later years, as he rose rapidly to be that firm's Chartering Clerk - a shipping broker - by 1890. This work involved matching up the import / export cargo requirements of a customer with a shipping company and all the paperwork involved. Only after he had proved himself did his father give him a one room office adjoining his own in Sandhill. Then in 1892, at age twenty-five, Arthur managed to persuade his wealthy father to raise the £22,700 necessary to build a ship and set up a shipping company. This was a completely new venture for his father. That first ship, the "Sutherland" – the first of seven of that name – was constructed by W. Dobson & Co. of Newcastle.

BJ Sutherland & Co. Ltd.

BJ Sutherland house flag.

The original company was called The Sutherland Steamship Co. with Arthur as its Managing Director and its fleet being composed of one new and two second-hand ships.

This became a limited liability company in 1902 under the new name of BJ Sutherland & Co. Ltd. with Arthur and one of his elder brothers, Benjamin John the younger, as joint Directors. Following the death of his brother 1909, Arthur was appointed Permanent Governing Director which post he held until his death in 1953.

This was the first of Arthur's three great ship owning ventures, the other two being the Tyne Tees and the Free Trade Wharf. Arthur named the company after his father who had died the previous year and he always retained that name in memory of him. By 1918 it had acquired several smaller fleets, all registered at the Sandhill address, which were managed under their separate company names of the Irismere Steamship Co., the Isles Steam Shipping Co. Ltd., the Grindon Steamship Co. Ltd., the Sandhill Shipping Co. Ltd. and the Dunrobin Shipping Co. Ltd. This was financial engineering to ensure survival if one or two went bankrupt and was also reflective of the transportation of different products. Arthur was gradually building up his shipping empire.

By 1920 all had been consolidated into the one company of BJ Sutherland & Co. which continued until his death in1953. In a 1944 letter to his youngest son Gordon, he made the proud statement that "This company has never missed paying a dividend and has regularly paid dividends varying from 5% to 10% over a great number of years". It was its success that funded his philanthropic ventures, and which underpinned the survival of the other companies in which he was involved.

His Will empowered his trustees to deal with the company after his death as they saw fit; their choice was to liquidate it.

Until the First World War his shipping interests were basically the export of coal and the import of grain but there were the beginnings of a burgeoning Scandinavian trade involving the export of coal and the import of ore and wood. The ore was needed to produce steel for his ships. He was adept at seeking new ports and cargoes, such as his being the first to import iron ore from Narvik.

""Arthur Munro Sutherland" by Arthur Kingsley Lawrence. Date unknown.

The wood was needed for both of his industrial interests; pit props for coal mining, and for the keel blocks, props, poppets (hull supports) and sliding ways needed to keep keels in place and cradle the hulls of his ships on the slipways. Between the Wars the range of produce carried expanded greatly.

Having inherited the management skills of his father and grandfather, an early stabilising factor in his life was that in 1893 he married into wealth. His wife was Fanny Linda Haggie, a daughter of his friend R. Hood Haggie of Haggies Rope Works then of Willington Quay "one of the largest rope manufacturers in the world". The Sutherlands and Haggies were family friends and were neighbours in Jesmond, the Haggies living around the corner at "Blythswood" in Osborne Road. Arthur had married the boss's daughter and Hood Haggie made Arthur a Director of his company as a wedding present. The Haggies also had interests in local politics, David Haggie having been Mayor of Gateshead in 1854. This could have been another factor in Arthur developing his political acumen, which included his father having been Sheriff of Newcastle.

William Doxford & Sons Ltd. of Sunderland. Turret deck steamers

Doxford were a family business of shipbuilders and marine engineers in Sunderland from 1837 and were always innovative in each of those areas. Their having the greatest output of any shipbuilder in the world in 1905 and 1907 for which the company was awarded honorary Blue Ribands, was not lost on Arthur and he and Sir Walter Runciman (who owned the Moor Line and later the Anchor Line) became major purchasers of Doxford vessels. Another factor in this was that Benjamin John the younger was chief agent for Doxford. Despite this, and somewhat surprisingly, Arthur was never a director of Doxford.

The second half of the nineteenth century saw many experimental designs in boiler design, engine configuration and the shift from iron to steel but none was more controversial than the so-called turret deck steamers.

The whaleback "SS Charles W Wetmore" departs from Detroit for Liverpool in 1891. Public Domain.

The docking of the "SS Charles W Wetmore", a Minnesotan grain carrier from the Great Lakes, at Liverpool was witnessed by Theodore Doxford in 1891. The vessel was named after the financier, Charles Wellington Wetmore who, along with John D Rockefeller, had financed the American Steel Barge Company to build whalebacks. It was broad beamed with little superstructure apart from a series of turrets to which minimal structures and winches were attached. The entire deck was cambered, in contrast to the standard vertical side of a ship between the gun whale and lower curved sections of the hull. Its shape broadly resembled that of a surfacing whale resulting in their being referred to as whalebacks. Theodore immediately saw the potential for better buoyancy and weight distribution. In due course, Doxford would connect those individual turrets within elongated trunks; the turrets were no longer visible, but the name turret deck remained.

In 1892 Arthur Haver, Doxford's chief draughtsman, modified the whaleback design into a single cargo space in the shape of two longitudinal cigar shapes running the length of the vessel. He also perceived that they had greater longitudinal and transverse strength and used less steelwork than equivalent vessels. Doxford applied for a patent that day under the title of "turret deck steamers" (albeit that the turrets were no longer visible!).

Arthur considered them to be "one of the most exclusive and interesting types of vessel ever produced on the North East Coast" and in 1893 purchased "SS Turret", the first turret deck steamer to be constructed for his then Sutherland Steamship Co.

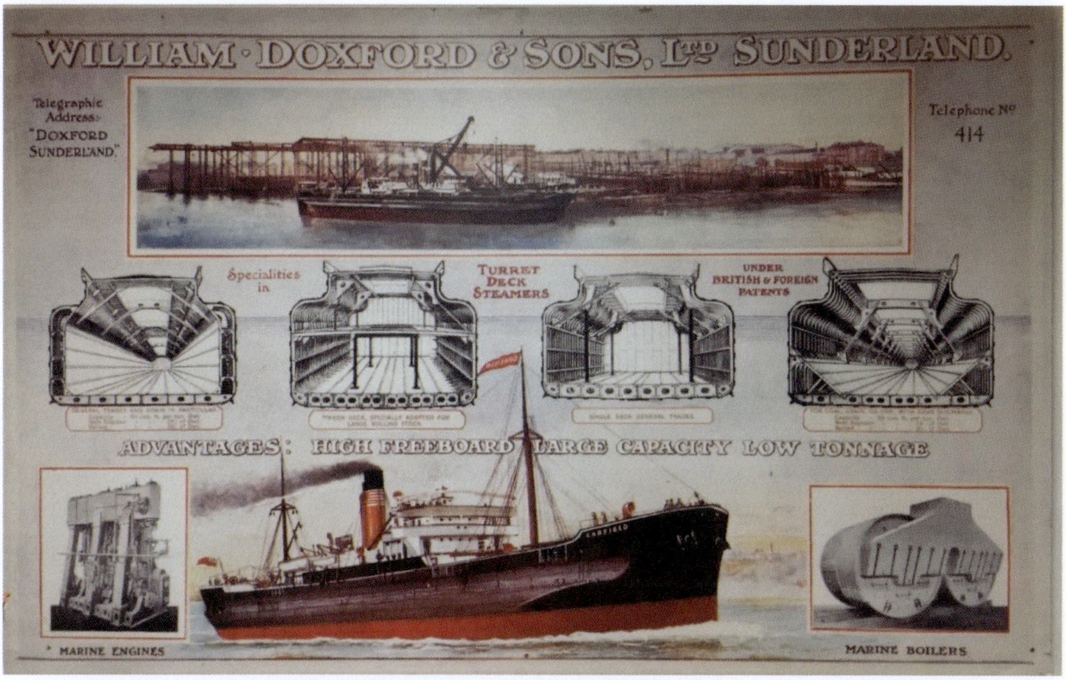

Doxford's advertising board for turret ships. By permission of Sunderland Maritime Heritage.

Many brokers considered them unorthodox and unwieldy and there was a Board of Trade enquiry in 1909 after three had capsized, which resulted in Doxford issuing special loading instructions. Nevertheless, Arthur and Benjamin John the younger pushed these forward whilst others hesitated because of their apparent ugliness and difficulties in manoeuvring. In the end Doxford would build one hundred and eighty-two turret deck steamers of which eight capsized, a loss-rate no worse than that of other ship types. In fact, the deck hatches of these ships were better protected from the ingress of storm waters than other designs. BJ Sutherland & Co. owned nine new and a further thirteen purchased from other shipping companies. The upper deck area was reduced because of those holds which, fortuitously, allowed lower fees when using the Suez Canal as the Canal charged on the breadth of the upper deck, and not the ship's beam. Many companies purchased these vessels for the sole purpose of avoiding Suez Canal charges though Arthur always denied that this applied to him. In fact, his ships travelling to the Far East tended to sail round the Cape of Good Hope as many of his stokers had died in the intense heat of the Red Sea. Fortuitously, this enabled him to set up new trading links with South Africa.

A Baltic escape

"Tynesider", Sir Arthur's biography, is in the main snapshots of his national and international commercial enterprises and philanthropic contributions. Some of the events in his life are given surprisingly short entries including his part played in this Baltic escape which does not warrant a mention in the book's index and only takes up one page in his chapter entitled "Scandinavia". It remains a largely unknown episode of the struggles at sea during the First World War but has finally been given due credit by naval historian David Parry.

Sweden, Norway and Denmark declared neutrality at the beginning of the First World War, but it was a desperate balance for them as the belligerents saw the strategic significance of their nations particularly after the Russian Revolution. Theoretically they could trade with whom they chose but the various belligerents went to great pains to close all trade except their own. Further tensions stemmed from the German and Russian navies being involved in hostile activities in the Baltic Sea up to the time of the Russian Revolution, with the British Navy being involved to a lesser extent via submarine warfare.

Ships caught in transit or in belligerents' ports at the outbreak of the First World War were, in general, released under the rules of the Hague Convention of 1907 and most captains took immediate action to get out of German ports within the few days of grace allowed. But for reasons unknown, ninety-two British and six Allied mercantile ships (sources disagree as to the exact numbers) chose not to take advantage of this arrangement and became trapped in twenty-one Baltic ports in Sweden, Finland and Russia by German minefields at the entrance channels. These ships represented three hundred thousand tons of additional carrying capacity for the British mercantile marine. Most were North East based though only three were owned by Arthur. Financial considerations were always paramount to him and he had to balance high insurance rates with, initially, low freight rates, and so no immediate action was taken.

But events moved rapidly when Germany introduced unrestricted submarine warfare from February 1915 and British merchant shipping losses increased alarmingly. Arthur's import / export links with Sweden would soon prove to be to his advantage, particularly his personal friendship with Dan Broström, President of the Swedish Shipowners Association and a long-term business associate; many of his ships had been constructed in Tyne and Wear yards. Arthur formed a Syndicate with the Swedes to enable the ships to get out. This was no loose arrangement but involved contracts with high rates to cover all the expenses for the mandatory Swedish captains and crews, plus an extra payment of £1,000 per vessel to the Syndicate, from which both Arthur and the Swedes benefited. It cost a pretty penny for Allied ship-owners to get their vessels out, but the Syndicate also needed to guard against potential bribery of Swedish captains who could have earned much more by informing Germany of the agreement. This arrangement was a purely commercial one.

In 1916 Swedish captains and crews got twenty-nine ships to Gothenburg often with help from Swedish torpedo boats that intervened whenever there was German harassment close to their territorial waters. It was a dangerous coastline with three-mile channels at the best of times plus rocks and minefields to negotiate. From Gothenburg the Allied crews took over and sailed the relatively short distance north along the Swedish coast, then hugged the Norwegian coast and finally headed west to the UK.

The Syndicate's operation was altered abruptly in 1917 following the Allies imposition of an embargo on Sweden due to its trade with Germany and its possibly re-exporting Allied goods to Germany. Swedish ships were not allowed to leave British ports resulting in a shortage of coal and grain, precipitating food riots in Sweden. The Swedes retaliated by laying mines and thus closing the Kogrund Passage in the Oresund (the narrowest passage into the Baltic). Questions were asked in Parliament of the Secretary of State for Foreign Affairs and letters were exchanged between the Royal Houses of Britain and Sweden. The Foreign Office and Admiralty then became involved with Arthur's Syndicate. Diplomacy won the day and Sweden re-opened the Kogrund. Up to fifty-four ships got out (certainly thirty-five in 1917 and sixteen in 1918) in exchange for the release of Swedish ships from British custody. The release of Swedish ships was politically significant to offset the possibility of Sweden giving up its neutrality and aligning itself with Germany. These were small but significant numbers of ships given the time needed to construct an equivalent number. Arthur reckoned that the carrying capacity of the ships restored to trade was 380,000 tons with a value of £5,000,000 with the cost of the venture being little over £300,000.

2 BALTIC EXTRACT.

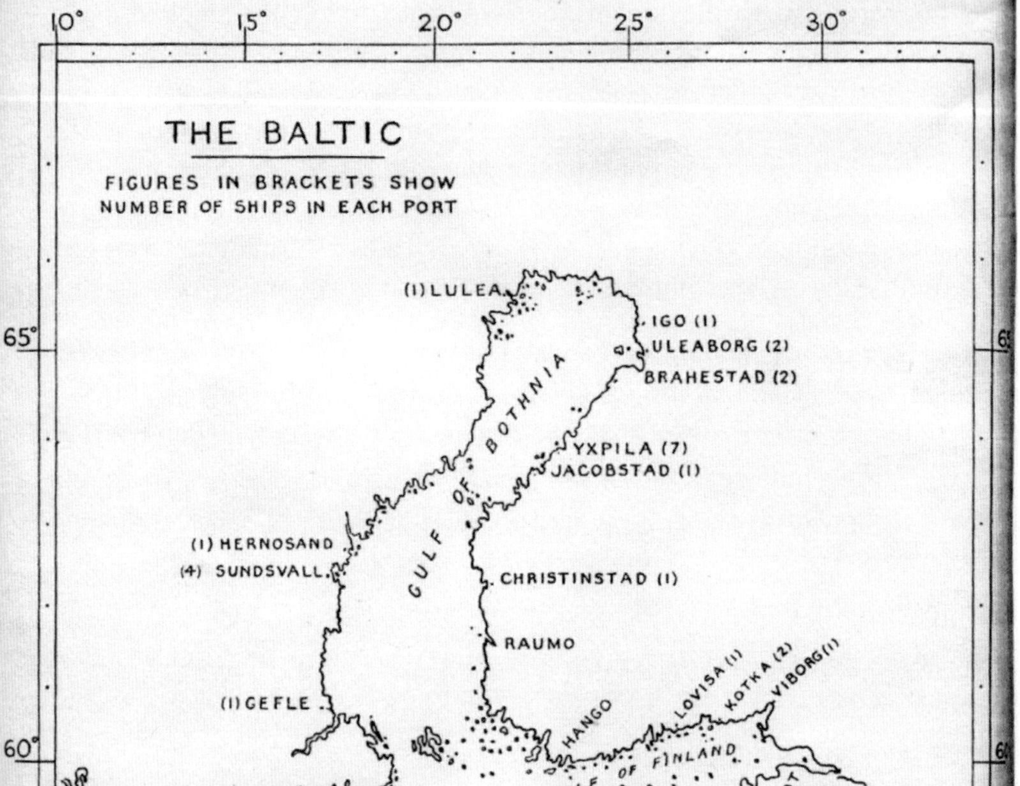

Baltic Escape ship locations in the "Baltic Extract page 2. 1939". Under licence from Tyne & Wear Archives.

The dilemma throughout these episodes was that Sweden was neutral and had to be seen to be so by Germany. Sweden was entitled to trade with both Britain and Germany so there were thin dividing lines.

The key to the success of this Baltic Escape was the friendship of Arthur with his former Swedish commercial partners, particularly with Dan Broström. Just before the outbreak of war, Broström, as an MP, became Sea Defence Minister whose portfolio included The Swedish Pilotage Authority which controlled mercantile pilotage. This was a serious dilemma for him as he could not be seen to be acting favourably to the UK. Consequently, the Swedish side of the Syndicate was fronted by Captain Janson (his works manager) and Ivar Lignell another Gothenburg ship-owner and insurer. This was such a sensitive political issue that it seems that the first public recognition of it was in 1936 when Sir Arthur was given the Honorary Freedom of Newcastle.

Broström was the catalyst in this escape. He was Chairman and Founder of numerous shipping lines, President of the Swedish Shipowners, an MP and a City Councillor and philanthropist to his hometown of Gothenburg. A posthumous portrait of him by the French artist Philippe de Rougemont (who specialised initially in Swedish Royalty, aristocracy and businessmen) captures his significance. He and his companies donated over one million Swedish Krona toward the founding of the Gothenburg Maritime Museum where his portrait still hangs. He was killed in a car accident in northern Sweden in 1925.

Dan Broström statue outside Gothenburg Maritime Museum. 1930.

Wikimedia Sverige. Connected Open Heritage.

His funeral service was held in Gothenburg's Lutheran Cathedral. His grandson, of the same name, is a venture capitalist in Singapore (2020) and still has the large silver tray in his office that Sir Arthur presented to his grandfather in recognition of Swedish co-operation in the Baltic Escape. Broström is an important historical figure in Sweden but his career would have ended in ignominy if his part in this Baltic escape had been known in his lifetime.

Posthumous portrait of Dan Broström by Philippe de Rougemont.

1931. Image by permission of the Gothenburg Maritime Museum.

A fortune made

Sir Arthur made his fortune immediately after the First World War when there was a severe shortage of available tonnage in Europe following the loss of many ships. Post-War optimism described as "a collective commercial madness" was not shared by him and he sold his entire fleet at the highly inflated price of £2.25million in 1920 to the Western Counties Shipping Co. Ltd. of Cardiff. Sir Walter Runciman (later 1st Baron Runciman 1933) the other major Methodist Scots / Tyneside ship owner at the time, had made a similar sale of £1.5million a few months before. Within a year the Western Counties Co. of Cardiff had gone into liquidation and Sir Arthur and Runciman bought back their own ships at a knocked down price during a market slump. They repossessed those that had not been paid for, and then resold them at current prices. Thus, their fortunes had been made judiciously by their selling when market prices were at their best and buying back during a market slump. Judging the best time to buy and to sell was an underpinning factor throughout Sir Arthur's business career.

Part of this fortune he invested into land in College Valley in north Northumberland, and farmland and coastland adjacent to Dunstanburgh Castle.

Between the Wars he traded worldwide – India, South America, British Columbia, Australia, The Mediterranean and South Africa - in almost every commodity with emphases on the export of coal and coke and the import of timber. Essentially, it was a tramp ship company which traded worldwide. (Tramp steamers, as opposed to scheduled ocean liners, carried anything in bulk such as coal, grain or ore to any port depending where the ships were and local economic conditions.) "The Ships List" gives details of seventy-six ships as a conservative estimate of the number of ships owned by the company (new or second-hand). The vast majority were named after Scots towns or counties and all bar one were constructed in North East, Clyde and Forth yards.

The Great Depression

Sea trade continued throughout the 1920s and 1930s, but many shipyards were concerned almost solely with ship repair as there was little confidence to invest in new ships. Doxford reached a record of launching a ship every fortnight before the First World War but were forced to close from 1924 to 1927 and had apparently launched their last ship in 1930. Nevertheless, the company set about designing a cost saving tramp ship – the diesel-engined Doxford Economy Cargo Ship of low capital cost and economical running costs. However, it seemed that it would never get off the drawing board. Then early in 1934 Sir Arthur commissioned Doxford to build him a modern fleet of those Ships which brought the Depression to an end in the North East. At age sixty-eight he had embarked on one of the largest shipbuilding programmes he ever undertook as he considered that the Depression had hit rock bottom though that was not then the commonly held view. He committed £1million – an unprecedented amount at that time - to Doxford. Seventeen were ordered. These were the first ships to be constructed on the Wear in four years. The Sunderland press proclaimed at the launch of the first just before Christmas on 16 December 1934 that, "Santa Claus has appeared in the form of Sir Arthur

Munro Sutherland" and "He drove silence and unemployment from the yards." These were highly efficient cargo ships which had not got past the drawing board until Sir Arthur recognised their worth. They would soon be acknowledged as setting new world standards for crew accommodation as recognised by Parliament. Sir Arthur had taken the risk with the launching of the first, the "Sutherland", and then Runciman followed suit. Once again, the two Methodist Tyneside entrepreneurs had blazed the trail. Sir Arthur claimed, and it is generally accepted, that he thus initiated the beginning of the end of the Great Depression in the North East.

There was a humanity within his capitalism and memories from his fleet sailors are encapsulated in the comment; "his fleet was well found, maintained and fed" as submitted by Stan Walton of British Columbia to "The Ships List" blog. Sir Arthur reckoned to offer fair wages, generated employment in a safe environment, and took opportunities to modernise his fleet in times of depression which he knew would end. He always made the point that many capitalists in shipbuilding and coal, such as himself, continued at a loss and without profit throughout the Great Depression when they could easily have sacked all their workers.

That Sir Arthur remained a Scot at heart as in his naming of some vessels after Scots towns and counties such as "Ross", "Inverness", "Dumfries", "Fife" and "Isle of Jura". He did not forget his ancestral counties of "Caithness" and "Sutherland"; seven were named "Sutherland".

Tyne-Tees Shipping Co. Ltd.

Tyne Tees Steam Shipping Company house flags.

This was founded in 1900 following numerous amalgamations and would become the second of his three great ship owning ventures. Unfortunately, few artefacts remain on the Quayside from those great days and even fewer relate to Sir Arthur. From 1919 he was a Director, and from 1921 -1943 he was Chairman of the Board of Directors. The future of the Tyne-Tees had looked bleak after the First World War in which half of its fleet had been lost and there was a shortage of available ships to purchase. Its ticket office (which was the original company office from 1864) was at 25 King Street under the Tyne Bridge and is now (2020) Sabatini's Italian Restaurant.

Sabatini's Restaurant, Quayside. Former ticket office for the Tyne Tees Shipping Company. Photograph by Nigel McMurray.

The passenger terminal was adjacent to the former Co-operative Wholesale building on the Quayside and is now the Malmaison Hotel. Sir Arthur would turn its fortunes around by diversifying and expanding its activities. The coastal passenger services between the North East and London had special rates for football fans travelling in each direction with coaches laid on at all terminals to get them to the London stadia. Passenger services would soon include stops on the East Anglian coast for the Norfolk Broads and day trips to Southend-on-Sea, the latter part of London social life. Regular trips to Germany and Holland continued. Paralleling these passenger trips were the

Hotel du Vin. Side entrance. Photograph by Nigel McMurray.

more significant, lucrative, and regular continental shipping services for coal, grain, cast iron, steel and sand for glass. In addition, new ships were designed and purchased with faster cruising speeds. Its large administrative, fitting out and repair building is now the Hotel du Vin in Newcastle's City Road.

What was in his day the Malmo Wharf, on the riverside immediately south of City Road and the Hotel du Vin, has now been renamed the Mariners' Wharf and converted into luxury flats. The only other geographical, trade and faith marker of those great days of trade with Scandinavia is the Sailors' Bethel (with which there is no direct link with Sir Arthur) which was erected by the Church of England for the numerous Scandinavian Lutheran sailors and their families who worshipped there.

Sir Arthur opted out of the Tyne-Tees venture in December 1943 due to war losses and his prophecy that after the War its viability would wane, under pressure from land transport systems coupled with Dutch competition on the North Sea and Channel routes. He monitored its share values over the years from the initial value of £1 rising to 23/- [£1.15] in 1923 to £3 in 1944. Consequently, he had sold the company when share prices were at the top of the market and his shareholders got a good deal. Coast Lines Ltd. took a controlling interest after his departure.

There were two company "house flags" which reflected the Newcastle and Scots loyalties of Sir Arthur. One featured the crest of Newcastle upon Tyne and the other the Scots saltire.

Free Trade Wharf Co. Ltd

This, on the Thames, was Sir Arthur's third and final great ship owning venture. The company owned only eight ships in its lifetime compared to the Tyne Tees which owned *circa* seventy and BJ Sutherland & Co. which owned *circa* seventy-six. However, he saw its primary purpose as being to maintain high quality wharfage on the Thames for the Tyne-Tees Shipping Co. and, significantly, for other cargo and liner companies which were charged appropriately.

Free Trade Wharf entrance, now luxury residences within a vast new build on The Highway, E1. 2020. Photograph by Peter Butt.

From 1884 the old Tyne Steam Shipping Co. Ltd. (one of the precursors of the Tyne-Tees) had only two steamer berths at its London terminal, the Free Trade Wharf on the Thames. From 1900 this became the Free Trade Wharf Co. Ltd., an independently managed company wholly owned by the new Tyne-Tees Shipping Company. ("Free trade" related to the period after tariff reform and the liberalisation of trade and this was celebrated with the naming of wharfs, halls and inns; an example in Newcastle is the Free Trade Inn on the Ouseburn.) Sir Arthur was Chairman of the Directors from 1921 to 1932 and under him it expanded into one of the largest river wharfs in London and was located at 14-20 Brook Street (now The Highway of the A13 from Essex into the heart of London) in Ratcliffe (or Ratcliff), London E1.

The name "Free Trade Wharf" is virtually all that remains of the riverside hamlet of Ratcliffe from which the great navigators of the sixteenth century set sail. The sole remnant consists of a fine gateway with two lions and the arms of the East India Company which was the original owner. This still leads to a long courtyard of fine warehouses which now constitute the more elegant conversions to luxury flats. Surrounding this courtyard are two enormous multi-tiered stacks of red-brick apartments which have replaced the vast complex of wharfs, warehouses, and yards of Sir Arthur's day.

Soon after Sir Arthur's take-over, six adjacent wharfs were purchased. Five new steamer berths and jetties with nineteen modern river cranes were built. The river was dredged to enable discharging at any state of the tide. New warehouses with a capacity of four million cubic feet were constructed and connected by new roads on a one-way system, all supplied with fresh water by a newly sunk artesian well. From 1929, he continued his principle of diversification with the company becoming ship owners. The design and construction works were undertaken by London companies and Brims of Newcastle (now Brims Construction Ltd, Sunderland). Ferro-concrete floors were laid for the storage of tea, canned goods and rubber, the latter with a Rubber Sampling Floor. Sir Arthur was Chairman of the Board of Directors which was composed of Newcastle and London businessmen. The principal import was "sea-borne coal" which arrived from the North East of England and South Wales. Sir Arthur's colliers were one of many fleets of independent colliers that came into the Port of London with most of the coal going to public utility undertakings (gas works and electricity supply companies), The London County Council tramways department, London Underground, cement works, sugar refineries, paper mills and the numerous industries on the banks of the Thames. Whereas coal for domestic consumption came mainly by rail from Yorkshire, coal could be transported to riverside plants cheaper by sea than by rail. A new administrative complex was opened on 31 May 1932 by the Mayor of Stepney and was named Sutherland House. The detailed brochure of the event was printed in Gateshead by his own Northumberland Press. The Board Room was a replica of his Chairman's Room in 40 Sandhill which was itself a replica of the 1658 oak panelling in Bessie Surtees House, so making it a little bit of the Newcastle Quayside in the East End of London. The oak panelling, office furniture, board room table and chairs were installed by two Newcastle carpentry companies; R. Rutherford and Robson & Sons. The Board Room was overlooked by his favourite portrait of himself next to one of his long-case clocks which gave the impression that he was a distant yet omnipresent and controlling figure.

Barnett Brothers, a privately-owned City of London ship-broker, also in Broad Street, dealt in the main with Scandinavian cargoes and looked after his interests in his absence; Sir Arthur was one of its Directors and so all angles were covered. As a private company it had no shares to buy or sell; Sir Arthur gave his company shares "as a present" in his Will to its managers who had served him well.

Sutherland House, Free Trade Wharf. 1932. By permission of the Northumberland Press Ltd.

Whenever Newcastle United were playing in London, Sir Arthur's passenger ships would discharge the fans at these wharfs which, coupled with the immense cargo trade regardless of tide, resulted in the Cockney stevedores referring to the Free Trade Wharf as "the madhouse".

Sutherland House board room. 1932. By permission of the Northumberland Press Ltd.

There was then almost a continuous line of wharfs on both banks of the Thames downriver from London Bridge to the Isle of Dogs and Greenwich, and the Free Trade Wharf ranked in the top ten of those wharfs. But it was derelict by 1970 and then the docks began to close from 1980 to be replaced by regeneration schemes including Canary Wharf and the Olympic Park. It was demolished in 1984 leaving only the main entrance, two warehouses and the original Newcastle Jetty within a modern luxury flat complex. The name Free Trade Wharf still appears in the London A-Z but now refers to the concierge-controlled luxury apartments in regenerated Docklands whose residents have no concept of the former usage of the docks on which their property stands.

"The madhouse." Tyne Tees Steam Ship Company Steamers off Free Trade Wharf in 1931.

"The Port of London Authority Monthly." June 1931

Free Trade Wharf [red brick] residences in 2020.
Photograph by permission of Brian Denton www.timecapturer.com

Honours in the world of shipping

Arthur was never content to be in control of shipping companies alone. His involvement was much greater than that in that he was always looking to improve the industry locally, nationally and internationally. This was recognised in his obituary in the "Shipbuilder and Marine Engine-Builder" of which this is a precis;

> "The death of Sir Arthur Munro Sutherland has removed an outstanding figure in British shipping and a shipowner of whom North-East Coast shipbuilders acknowledge their debt.
>
> In the depression he gave orders for ships that had evolved but had not proceeded beyond the drawing board. He never lacked confidence. Sir Arthur was a former President of the Chamber of Shipping of the UK, Past-President of the Newcastle and Gateshead Chamber of Commerce and Past-Chairman of the Tyne Improvement Commission."

His monumental skills and foresight were recognised at an early stage in that by 1913 he was Chairman of the North of England Steamship Owners Association. By 1930 he was elected as President of the Chamber of Shipping of the UK, "the highest honour of the [British] shipping world". The Chamber's function as described in its current website mirrors that since its foundation viz. "to be the voice of the UK shipping industry and to work with Government, Parliament and international organisations on behalf of our Members. To develop and implement a national strategy and give, advice information and expertise." The Chamber considered its role in the pre-Second World War era as having a "vital function in the conduct of the county's most important industry".

However, that honour would be exceeded by his 1938-1950 Presidency of the Shipping Federation. This had been founded in 1890 to promote the interests of shipowners and as a "fighting machine" to counter the increasing power of the unions, in particular those led by the Sunderland-born temperance leader James Havelock Wilson. But successful national strikes in 1911 had obliged the Shipping Federation to adopt a more gracious stance. To everyone's surprise, Sir Walter Runciman recognised Havelock Wilson's National Union of Seamen, and the beginnings of the Shipping Federation as an arbitration organisation had begun on Tyneside. Runciman was its first President followed by Sir Arthur. It involved the overseeing of the twenty-three Districts of the Federation and presiding over the one-hundred-and-forty-member Executive Council which met in London twice a year. This was a grouping of management and unions and was a remarkable achievement between two diametrically opposed groups. Perhaps in their Methodist sub-conscious, Runciman and Sir Arthur recalled Wesley's edict to be "the friends of all and the enemies of none."

Sir Arthur chairs a meeting of the Tyne Improvement Commission. Under licence from Tyne & Wear Archives.

The navigable section of the Tyne is a man-made river and since the founding of the Tyne Improvement Commission in 1850, with its headquarters in Bewick Street, it has been responsible for its dredging, straightening and the protection of the Tyne's entrance. Arthur was a Member from 1911, a very early recognition of his love of the Tyne and his intention to keep it at the forefront of world shipbuilding. He was Chairman from 1935-1945. The new deep-water quay at Tyne Dock, constructed during his Chairmanship, was named after him as Sutherland Quay. Many derelict and unoccupied sites on the Jarrow riverside were cleared during his time as Chairman. (The Commission was dissolved in 1968 and was replaced by the Port of Tyne Authority.) His involvement in Tyneside shipping never waned and he was the natural choice for Chairman of the Tyne Port Emergency Committee in the Second World War which met every Monday morning at his Sandhill offices.

He still loved being involved in the nitty-gritty as reflected in his still being President of the North of England Steamship Owners Association in 1953, the year of his death, by which time he was known as "the father of the Quayside".

Nationally he was a Committee Member of Lloyd's Register of Shipping with its safety functions and distribution of its Daily Lists giving details as to what was going in and out of every British port. He always had his finger on the pulse of shipbuilding and trade. He visited the USA over thirty times on business matters. This included his being a British Representative of the International Maritime Conferences. He always kept a weather eye on activities in the USA which he always saw as the UK's potentially greatest competitor.

He never believed in luck and attributed his success to recognising opportunities. He made rapid decisions while others hesitated. When he was in the chair, business was concluded in record time. He disliked protracted discussions on anything he considered to be a waste of time. Sir Arthur was always aware of innovation in the shipping industry, was a pragmatic capitalist, prepared to take on what to others were huge risks, had an unshakeable confidence in Geordie skills and tenacity, and was always an optimist. He was a workaholic as in "work is my hobby". The motto on his [coat of] arms said it all; "Without fear".

Shipyards

Sir Arthur was first and foremost a steamship owner and shipping-broker. He funded the construction of ships but owned few shipyards. However, he was directly involved with the managing of two shipyards, one in Southampton and the other in Blyth. His management skills were brought to these companies on the recommendation of close friends and colleagues in the shipping trade. His involvement in Southampton was primarily with luxury yachts, quite a departure from his knowledge base. His involvement at Blyth almost proved disastrous to his career and reputation.

White Brothers of Southampton

This company, established in 1892, were yacht builders which role they still fulfil today as the builders of luxury sailing yachts. Sir Arthur was a Director alongside Sir Walter Runciman, Sir Herbert Cory MP, John Cory the younger and AR White, the latter a descendant of the original White Brothers. The Corys, a West Country family, had a history of ship building and trade. All five were Methodists. Sir Arthur and Runciman were two of the most prominent Tyneside ship owners with Runciman's primary company being the Moor Line. They were personal friends, Methodists (Runciman was a Methodist Local Preacher) and both had a positive attitude to trade unions. Unlike Sir Arthur, Runciman was a Master Mariner and a yachtsman of note, being a member of five yacht clubs including being a founder member of the Royal Northumberland Yacht Club and hence his interest in Southampton and environs as an international sailing centre. The Corys had purchased the original company after it went into liquidation in 1923 and, presumably, Runciman then brought Sir Arthur onto the board for his financial and managerial expertise. It was then a relatively small company with each share purchase price being only 10/- [£0.50] but its new Directors saw potential. Sir Arthur made a point of always attending committee meetings of those companies of which he was Director and presumably this was achieved when he and Runciman were in Southampton for Shipping Federation meetings.

White Brothers share value collapsed to 6d [£0.03] at the outbreak of the Second World War. The company had never paid a dividend. However, Government contracts and naval work saw it through the War during which its shares rose to 18/- [£0.90] per share at which point (1944) Sir Arthur sold it! He had anticipated its collapse after the War but, on this occasion, his prophecy was incorrect in that after 1945 its repair yards blossomed, and it moved successfully into the luxury yachts market.

Blyth Shipbuilding & Dry Docks Co. Ltd.

Ships were first built on the south bank of the River Blyth in Northumberland in 1811 and until the Second World War Blyth was Europe's largest coal exporting port. This company, first registered in 1883, had a chequered history and carried out its trade under various names until its closure in 1925. It was resurrected the next year by Robert Stanley Dalgleish to be followed by a further closure in 1930. Dalgleish re-opened it again in 1937 with Sir Arthur in charge as Lead Director. Dalgleish, as Managing Director, ran the company on a day to day basis and was Sir Arthur's deputy. Walter Runciman (now Lord Runciman) and his son of the same name were soon appointed as

Directors. Dalgleish and Sir Arthur had been close friends in relation to the Tyne for decades and both were former Lord Mayors of Newcastle, but this was the first, and last, time that they were business partners. The company was a significant ship building facility with six berths and five dry docks. It flourished during the Second World War when it built frigates, corvettes and mine-sweepers for the Admiralty. It was never a profitable company until it received war time contracts; its £1 shares were valued at 1/- [£0.05] at its lowest ebb but these rose to 37/6 [£1.87] in 1944 at which point Sir Arthur sold it. However, its heyday came after the War when under different owners but at the same location it constructed tankers, ore-carriers and coastal steamers. It closed for the last time in 1967.

The Blyth Docks Bribery Case

There was always a struggle between the rival shipbuilders during the War to obtain lucrative Admiralty contracts in which regard the Admiralty was always alert to possible malpractice. In 1943, as the Lead Director of the Blyth Dry Docks and Shipbuilding Company, Sir Arthur was charged, along with Robert Stanley Dalgleish, the Managing Director, with conspiracy on five counts of giving bribes totalling £155 to Charles James Butt an Admiralty official. An eyewitness, then a milk delivery boy, recalled seeing simultaneous police raids at 6am on the (adjacent) homes of Sutherland and Dalgleish and their removal in police cars. They were then charged with insider trading but were not detained. The accusation was that Blyth Docks had unlawfully obtained information on a contract for two minesweepers. Blyth Docks had already won the tender, but the accusation was that on obtaining inside information from Butt, Blyth Docks then increased their tender from £61,000 to £67,000 though ensuring that it was still the lowest tender. Sir Arthur had signed the amended tender. Charles Wintersgill, the former Secretary and Commercial Manager of the company, was the main prosecution witness; he had been previously employed at Blyth Docks but had been dismissed by them for theft and was in Durham Gaol. Sir Arthur immediately perceived the gravity of the situation and realised that a guilty verdict would result in him losing his civic rights as well as his credibility in society, trades and Methodism and his whole world would collapse. He needed to choose his counsel well and chose to be represented by Sir Patrick Hastings KC (who came out of retirement to represent Sir Arthur) and Dr John Charlesworth. Hastings had been Labour MP for Wallsend from 1922-6 and was Attorney General in the 1923 Labour Government. He was "one of the outstandingly brilliant advocates of his day and generation" and of "possessing the power of deadly cross-examination". Charlesworth was a Newcastle barrister and joint author of the definitive "Charlesworth and Percy on Negligence".

Sir Arthur and Dalgleish were both former Lord Mayors of Newcastle, each owned their own companies in Sandhill, lived next door to each other in Fernwood Road ("South End", Dalgleish' vast red brick house is still extant) and both had held office simultaneously in the Newcastle Commercial Exchange and Tyne Improvement Commission. Just a few months prior to charges being laid against them, Sir Arthur had presented Dalgleish with a parchment commemorating Dalgleish' fifty-year association with the Quayside.

The initial hearing of three days was held in Blyth Police Court at which the local press commented that "Dalgleish had a uniformed female nurse in a white cap in constant attendance" and that "all three defendants brought their own cushions". This may sound amusing now but it was part and parcel of Hastings intentions of gaining sympathy for the defendants. The case for the Crown was led by C. Paley Scott KC. It was widely considered that Hastings would get the case stopped in its tracks, but he failed in this and the case was remitted by the Blyth magistrates to Northumberland Assizes. However, given that Sir Arthur was High Sheriff of Northumberland and thus accompanied the Judges to the Assize and empanelled the juries, the case could not be heard in Northumberland. A solution would have been to commit the case to the Newcastle Assize. However, Sir Arthur was a magistrate in Newcastle and thus a member of the Newcastle judiciary. And Butt, the Admiralty official, had instructed Swinburn G Wilson as his solicitor; Wilson was Under Sheriff of Newcastle. Therefore, it was decided to hold the case at Leeds Assizes where there was a five-day trial.

Dalgleish and Butt were found guilty on all charges. Dalgleish was fined £500 and imprisoned in Wakefield Gaol for fifteen months. He died in 1944, soon after his release. Butt was sentenced to three years penal servitude. Sir Arthur claimed throughout the trial that he had been entirely unaware of the actions of Dalgleish as Managing Director and had signed the tender amendments in good faith. Hastings duly shredded Charles Wintersgill, the Prosecution's chief witness, on the grounds that he was unreliable in that he had been gaoled for theft from the company. Sir Arthur was found not guilty on all charges and was acquitted.

Sir Patrick Hastings KC. Under licence from Getty Images.

The personal consequences to the Dalgleish and Sutherland families of the different outcomes were severe. Three years later, Sir Arthur, in "Tynesider" his autobiography, gave a glowing account of Dalgleish albeit laced with comments on his personality such as "irrepressible", "stormy petrel" and "his impetuosity sometimes recoiled on his own head". He made no mention of the Blyth Docks Bribery Case.

A strange sequel to this was reported in the Sunday Express of 1st August 1943 under the heading of "£1,000,000 offer to save man from jail". The reporter referred back to the trial and that "During the lunch adjournment, Sir Arthur asked me 'What do you think will happen?' I said I was certain he would be acquitted and his friend [Dalgleish] would go to jail. Sir Arthur replied; 'I am ready to pay £1,000,000 to stop my old friend from going to prison. He has a blank cheque from me'. He was shocked when I pointed out that there was only one sentence for conspiracy – imprisonment. He made me promise 'not to write the story until my Will is published'. When Alderman Dalgleish' property was dispersed it included the blank cheque from Sir Arthur." The Sutherland family consider this to be a viable story.

Trade unions and socialism

As a Wesleyan Methodist Sir Arthur tended towards philanthropy rather than socialism. His speech on "Industry, Thrift and Ambition" when he received the Honorary Freedom of the City of Newcastle said it all; "Don't be led away by the belief that Government aid can ever adequately substitute individual effort. Thrift means turning off the taps of waste resulting in something which will be for the good of the whole community. Ambition is to be the 'boss' and not the servant, to be a landlord and not a tenant, to be of service to our fellow citizens."

In his "Toast to the Shipping Industry" at the Annual Banquet of the Chamber of Shipping in 1930 he emphasized his belief that capitalism, not socialism, was the answer even in a depression.

> "We are going through a period of acute Depression, but it will only be transient. Depression is not decay. All storms, commercial as well as atmospheric, rock themselves to rest. There is no need for the Government to supply shipowners with money to buy ships, but if someone will whisper in the ear of the Chancellor that shipowners are really philanthropists and therefore should be free of all taxation, the

money saved could be used to build up our businesses and would do more good to remedy unemployment than the pouring out of dole relief that is being done at the present time."

This contrasted starkly with the Primitive Methodists – those who had returned to "primitive" i.e. original Methodism - with their emphases on field preaching and radical politics. In many respects, Primitive Methodism and trade unionism can be regarded as parallel movements with each ministering to the needs of the so-called working-class population. Methodism sought the reform of the spirit of the individual and by so doing affected the whole of that person's life – political, social, industrial and religious. An analysis of eighty English trade union leaders in England in Sir Arthur's lifetime shows that seventy came from Methodism and forty-six from Primitive Methodism; most were preachers, and many were Sunday School teachers. Sir Arthur's late mother was a Primitive Methodist giving him a high regard for trade unionists which he would not otherwise have had. Sir Arthur did not cast aside his Primitive roots as seen in his remembering members of the Proud family (his mother's family) in his Will.

Trade unions and shipbuilding

One would have expected Sir Arthur's most formidable opponent to be Ernest "Ernie" Bevin the Dockers' leader and General Secretary of the Transport and General Workers Union but the two leaders were pragmatic enough to be able to work together – "Mr Bevin always does his best to get a good bargain for his workmen" and "he was blunt, forthright and absolutely sincere in his convictions"

Other Methodist trade union leaders, John Wilson and Mark Hodgson, offered him potential opposition in the boiler makers and steel trades of the shipping industry. But Sir Arthur's potentially most formidable opponent was John Hill of Govan, the General Secretary of the United Society of Boilermakers and Iron and Steel Shipbuilders, a charismatic leader of "Red Clydeside" (the radical and militant Labour movement in Glasgow in the 1910s and 1920s) and the 1917 President of the Trades Union Congress. Such was the industrial power of Newcastle that from 1890 it was home to the national headquarters of the Union and thousands, including the Lord Mayor of Newcastle, attended the opening of Lifton House in Jesmond (named after the Devon birthplace of Hill's predecessor), which became known as "one of the most famous union headquarters in the world". There would be no steam for the steamships without boilers and "hard battles in the interests of the boilermakers" were common. But to Sir Arthur, John Hill was but another "doughty and sporting opponent." "John Hill said what he thought and meant. Each month his circular ["Remarks" in the Boilermakers Monthly Report] tilted a lance at capitalism. His watchful pen castigated shipbuilders and shipowners alike. None did it better than John Hill".

"History of the United Society of Boilermakers" by DC Cummings. 1905. Front cover features Newcastle.

The expected clash of the Titans never materialised as the two of them had much in common. Faith issues were important to each; John Hill was also a non-conformist (he was a life- long Congregationalist) and in his biography, "Boilerman's Heid", his grandson states that Hill always reckoned that "The Bible is still my best book" which inspired him "to fight for a better society". Both were teetotallers, were Newcastle magistrates (Hill was a member of the Licensing Branch) and they lived within half a mile of each other. Sir Arthur had a remarkable ability to defuse potentially explosive situations and another factor in his favour in that he remained loyal to his origins as virtually all his ships were built by yards on the Tyne, Wear, Clyde and Forth.

Nevertheless, Sir Arthur always took a firm stance against strike action and nationalisation which he considered would always be counterproductive though surprisingly, he supported the British Shipping Assistance Bill of 1935. He had finally conceded that North East shipbuilders could not compete with overseas companies which were heavily subsidised by their governments. Indeed, he went as far as becoming a Member of the Tramp Shipping Subsidy Committee. His capitalist principles *could* be diluted if pragmatism demanded it.

Trade unions and coal

Millions of tons of coal were exported in his vessels from the North East. Sir Arthur was a Director of John Bowes & Partners Ltd, a major coal mining company in County Durham whose Chairman was Lord Glamis, the Queen's brother and he forever defended the mine owners against nationalisation. In 1936 Sir Arthur accompanied the Duke and Duchess of York (later King George VI and Queen Elizabeth) in their opening of the Jarrow Coal Staithes to serve the Duchess' Bowes Collieries. Thus, the Duchess, as a Bowes-Lyon and owner of the Bowes Collieries, was opening the staithes which would allow her coal to be deposited into Sir Arthur's ships. Some of Sir Arthur's Methodist trade union contemporaries in the coal fields included Tommy Burt (a fellow resident of Jesmond) and William Straker, both sometime General Secretaries of the Northumberland Miners Association. Also, Peter Lee the President of the Miners Federation of Great Britain. All four were Methodists. Negotiations were always as amicable as they could be given that they had "the steadfastness of pit-row chapel behind them." Sir Arthur had more respect for trade union leaders than may have been expected, but there again, most of them were Methodists like himself and his mother was a Primitive Methodist. Perhaps his mother's hand of restraint was always on his shoulder.

Relationships between employers and unions were *generally* amicable as reflected in the "Colliery Guardian" journal which covered every aspect of the coal and iron trades in Northern England. Everyone was aware as to the issues, whether it be Sir Arthur making his points against proposed nationalisation and strikes, to constant health and safety issues, to the illegal racing of pit ponies in the mines. Communications were surprisingly good though not enough to

William Straker. Newcastle Black Plaque. Photograph by Nigel McMurray.

eliminate all the tensions between employers and workers. Nevertheless, Sir Arthur was a Member of the Newcastle Methodist Coalfields Commission (which operated alongside the Methodist Coalfields Distress Fund) which was one of four such Commissions in the major coalfields of Newcastle, South Yorkshire, Notts & Derby and East Kent. They operated between 1916 and 1932. They were set up by the Wesleyan Methodist Conference of 1916 to relieve the distress and financial needs of miners in those areas via the distribution of moneys and the construction and extension of chapel halls. Despite his intense capitalism, Sir Arthur was prepared to offer both financial support and his financial acumen to miners' families in distress and the miners trusted him to do it.

As Sir Arthur looked out of his window overlooking the Quayside after the Second World War at his beloved 'Change he prophesied that Tyneside enterprise was nearly all over by 1947. He considered that decline had set in as early as the 1926 Coal Strike culminating in Nationalisation in 1946. His prophecy that Nationalisation would result in the disappearance of the recognised great brand names of British coal proved to be the case. Vesting day of the National Coal Board in January 1947 was a very dark day for him even though £164million compensation was paid to the former colliery owners including himself. He believed that industry must be unrestricted and unfettered if it was to progress.

Behind his grand, opulent appearance and self-confidence he had a remarkable ability not to antagonise those at the other end of the political spectrum. Surely this was reflective of a Wesley comment; "Be calm. Be good-natured. Be courteous. Be merciful – when you have gained an advantage over your opponent, do not press it to the uttermost."

Other business interests

Sir Arthur's business interests stretched way beyond shipbuilding.

Dunstanburgh Castle and Golf Course

Sir Arthur made two major land investments in Northumberland of which this – the farmland and coastland from Craster to Beadnell – is one. For the other, that of the College Valley near Wooler, see **HOMES Hethpool.**

"A cutter in distress to the south of Dunstanburgh Castle" by John Wilson Carmichael. 1865.

In 1919, he purchased Dunstanburgh Castle with its Arthurian legends. A fine oil painting of the Castle still hangs in the former games room of Thurso House - his town house and now the Mansion House. He purchased the entire estate whose glory had long since departed and was then under arable cultivation and consisted of all the farmland between Craster and the River Long Nanny near Beadnell. He carried out eight years of restoration on the Castle site and then, very astutely, retained the land but obtained a Deed of Guardianship with HM Office of Works to maintain it. His Will of 1953 left the Castle, two farms and the coastlands to Benjamin Ivan, his eldest son, who had managed the estate since 1945; he would dispose of it to English Heritage and the National Trust in 1960.

James Braid. WD & HO Wills Cigarette Card. c1930.

Based on a painting by Herbert "HH" Harris fl.1918-40.

The estate included the Dunstanburgh Castle Golf Course – golf being perhaps his only leisure time activity - and he employed James Braid, a premier British golfer and golf course architect, to redesign the course. Braid had won the British Open five times and then began designing one hundred and sixty golf courses including Carnoustie and Gleneagles. Braid completely remodelled the existing Dunstanburgh nine-hole course (now eighteen-hole). It has stood the test of time and describes itself today as follows; "with isolated beaches, historical fortifications and flora and fauna of colour and variety, there is rarely a moment where the horizon doesn't offer something to savour" with the rider that "the course plays according to the elements and if the weather turns against you so the course reveals considerably more ferocious teeth". Sir Arthur also had a vision to create an "Embleton Riviera" style resort with a wide crescent of twenty-three high-class villas, an extension of four such villas already established when he purchased the estate. This dream never came to fruition and only the original four remain, albeit various wooden bungalows 'appeared on the beaches' which were given official leases retrospectively.

Newcastle Chronicle and the Newcastle Aero Club

He owned the Newcastle Chronicle Group of seven newspapers for five years. He frequented the Union Club for lunch and on one such occasion in 1920 he emerged as owner of the Chronicle Group at a cost of £800,000 having purchased it from the family of Joseph Cowen over lunch. Sir Arthur always liked to joke in his after-dinner speeches that he "had just called into the Union Club to buy a paper". He appointed Munro, his eldest son, to manage the project. Apparently, it did not prove to be a sound investment as, even then, the new national presses began to make inroads into the territory of the provincial newspapers. Sir Arthur sold the Group after just five year to James Kemsley who would in due course become Viscount Kemsley, a newspaper magnate. Sir Arthur took a considerable financial hit from this premature sale.

The sale was to the great annoyance of Munro as he recollected that the workforce had been promised that Sutherland money would always be available for the business. Perhaps the real situation regarding this is only now being understood. Munro was a Flying Officer in the Royal Flying Corps' 29 (Fighter) Squadron during the First World War. He was forced down by enemy aircraft within five weeks of arriving in France and was a Prisoner of War for two years. Sir Arthur put him in charge of a major managerial and financial operation almost immediately after his return. The traumas and stresses of losing friends in combat and the humiliations associated with being shot down and imprisoned were not then understood. The difficulties with his father could have been triggered by the newspaper sale precipitating further stress from another apparent failure in his life. Munro moved out of the family circle and farmed in north Northumberland, a happily married man with three children, until his death in 1941. At least he was spared another tragedy; his son, Pilot Officer Peter Munro Sutherland followed in his father's footsteps and joined the RAF only to be killed when his aircraft crashed approaching Aberdeen Airport in a night navigational exercise in 1944.

Nevertheless, Munro, perhaps inadvertently, aided his father in his dream to make Newcastle "an airport city". In 1919, the Newcastle Corporation's attempt (with Sir Arthur as Lord Mayor) to convert the Town Moor into a civil aerodrome was blocked by the Freemen. In "Tynesider", his autobiography, Sir Arthur recollected having received a postcard from an irate citizen;

"To the Lord Mayor and Corporation with very few exceptions, Arthur Sutherland will be remembered as the Lord Mayor who tried to cheat Newcastle people out of their Town Moor: the lungs of the poor as well as the rich of Newcastle."

However, in 1925, Munro along with Baxter Ellis and Philip Forsyth Hepple (all former RFC pilots and Prisoners of War) pursued his dream when they founded the Newcastle Aero Club as the first flying club in the country. It operated initially from an airfield in Cramlington prior to moving to Woolsington in 1935 which coincided with the opening of Newcastle Airport at that location. The Club trained hundreds of pilots prior to going into liquidation in 2004.

Aston Martin

In its "Centenary Celebration Programme 1913-2013", Aston Martin claimed to be the "world's most charismatic car company" where "brilliant minds and dedicated craftspeople have created cars that have become the most coveted and iconic of all-time." However, without Sir Arthur's financial intervention in the 1930s, the brand would never have survived, and James Bond's "company car" would not have been an Aston Martin. They have won concours prizes at Pebble Beach Concours d 'Elegance, Louis Vuitton Classic, Cartier style et Luxe ("the finest concourse d 'elegance in the world"), Classic-Gala-Schwetzingen and NEC Birmingham Best of Show.

Gordon, Sir Arthur's youngest son, was a petrol head. He chose not to follow in his father's footsteps into the shipping business but became a first-rate engineer (obtaining a Diploma with Honours from the Chelsea Automobile Engineering Training College) with expertise in racing cars and in October 1932 he persuaded Sir Arthur to purchase the works in Feltham, Middlesex. This was deemed "one of the most important announcements in Aston Martin's history". It had stumbled through the Great Depression and it took Sir Arthur's investments to give the company a long sought after financial stability. Within three years Sir Arthur would become Chairman of the Directors with Gordon as Managing Director and Engine Developer. Gordon would race them at Brooklands including an appearance in the last ever race at the circuit which he lapped at 105.97mph. Sir Arthur and Lance Prideaux Brune, the London distributor for the company, bailed the company out in 1932. Sir Arthur injected a further £10,000 in 1935 when he became Chairman and carried out his usual radical review of management and supply chain procedures (including his doing away with many of the bespoke components and replacing them with those more easily obtained from other manufacturers) to make the company viable. Gordon's wife, Helen, joined the Board in 1937 subsequent to the acrimonious resignation of Enrico Bertelli whose company was a major supplier of chassis; Sir Arthur considered that Bertelli had too much influence on the Board and Bertelli reckoned that Gordon had insufficient experience in the motor trade.

Brooklands in the 1930s

TWELFTH RACE

DISTANCE: ABOUT 9 MILES. (FORK START.)

(Cars will run three complete laps and finish in the Railway Straight.)

5.55 p.m. **THE THIRD AUGUST OUTER CIRCUIT HANDICAP.** (The entrant of the winner to receive £25; the entrant of the second £10; and the entrant of the third £5. Two to start, or no race; four to start, or no second prize; seven to start, or no third prize.) For motor cars in racing or touring trim propelled by means of internal-combustion engines only. Entrance 3 Guineas. Closed 26th July, 1939.

No.	Entrant.	Car.	No. of Cyl.	Bore.	Stroke.	Cubic Capacity.	Driver.	Colour: Car	Wheels	Start. m. s.	No.
1	Capt. A. G. Miller ...	Lagonda ...	6	88·5	120·6	4,467	P. Nuttall ...	Blue :	Black ...	0 0	1
2	Mr. B. H. Talbot ...	Aston-Martin ...	4	78	102	1,950	Entrant ...	Grey :	Silver ...	0 21	2
3	Mr. C. G. H. Dunham	Alvis II ...	4	73	110	1,842	Entrant ...	Silver :	Blue ...	0 21	3
4	Mr. E. Winterbottom...	Alta ... S	4	60·5	95	1,092	Entrant ...	Black :	Brown ...	0 24	4
5	Miss D. Stanley-Turner	M.G. ... S	4	57	73	747	Entrant ...	White :	Black ...	0 24	5
6	Mr. F. R. Gerard ...	Riley ...	4	69	100	1,496	R. M. Turner ...	Blue :	Ivory ...	0 24	6
7	Mr. R. G. Sutherland...	Aston Martin ...	4	78	102	1,950	Entrant ...	Red :	Red ...	0 28	7
8	Mr. E. M. Thomas ...	Frazer-Nash B.M.W.	6	66	96	1,971	Mrs. E. M. Thomas	Black :	Black	0 32	8
9	Mr. J. H. R. Smith ...	Frazer-Nash S	4	69	100	1,496	Entrant ...	Blue & Silver :	Silver	0 42	9
10	Mr. G. L. Baker ...	Graham-Paige	8	88·6	114	5,387	Entrant (1) ...	Blue :	Blue ...	0 42	10
11	Mr. B. Burton ...	Talbot ...	6	75	112	2,970	Entrant (2) ...	Blue :	Silver ...	0 42	11
12	Mr. D. Creed ...	Lagonda S	4	62·5	90	1,104	Entrant ...	Silver & Red :	Silver	0 51	12
14	Mr. R. F. Oats ...	Alvis ...	4	68	103	1,496	Entrant ...	Red :	Black ...	1 5	14
15	Mr. R. C. K. Money ...	M.G. ...	6	57	71	1,087	Entrant (3) ...	Black :	Red ...	1 18	15
16	Mr. A. J. Welch ...	M.G. ...	4	63·5	102	1,292	Entrant ...	Blue :	Silver ...	1 28	16

S denotes supercharged engine.

Competitors in any previous handicap at this Meeting may be re-handicapped at the discretion of the Handicappers.

RESULT.

Winner 10 *Second* 11 *Third*............ *Speed of Winner* 99·46 *m.p.h.* *Won by* 1/5 second

Programme for the last race at Brooklands. By permission of the Brooklands Museum.

Helen and Gordon Sutherland with the 1932 Rudge-Whitworth cup. By permission of Dr. Tom Rollason.

Sir Arthur remained cautious regarding the risks of being too heavily involved in competitive racing but, nevertheless, authorised team appearances at the 24 Heures du Le Mans; Aston did not provide an individual winner in any of the classes but their three cars entered in 1932 and 1935 won them two Rudge-Whitworth Cups awarded to the best ranked marque. He also authorised racing in the Ulster Tourist Trophy. He then took the advice of Lord Ridley (3rd Viscount Ridley) of Northumberland, a car racing enthusiast who set various speed records for the flying kilometre and flying mile at Brooklands. On this advice, Sir Arthur invested £10,000 into racing engine development on the grounds that racing played

Fay Taylour. Pattreiouex Cigarette Card. 1930.

an important part in the development of a car as well as valuable publicity.

He was even persuaded to enter Fay Taylour, the premier woman motorcycle speedway rider, as co-driver in the one-off "Mussolini Gold Cup" of 1934, a 3,500-mile race around Italy in which she finished fourth. One of his most successful drivers was St John 'Jock' Horsfall who, in 1938, drove the 2-litre Speed Model to success in the Leinster Trophy and to second place in the RAC Trophy Tourist Race at Donnington Park beating the more fancied BMWs. Sir Arthur much preferred the gentler and less financial risk of rallies, trade events and shows. Somewhat surprisingly, he authorised elegant ladies and film stars to grace the already good-looking vehicles. Many privately sponsored racing drivers soon took to Astons; they were no drain on the Works team's strict budget.

Despite the international reputation of its cars for design and performance it always struggled financially (its Profit and Loss account showed a debit of £56,758 in 1939). Its survival was dependent on its being a subsidiary of Sir Arthur's highly successful BJ Sutherland & Co. which held shares in it. Thus, Aston Martin had not been self-sufficient even under Sir Arthur.

The "Atom" (a name chosen to represent something small but powerful) was the last hurrah for the Sutherlands in their adventure into the automobile industry. This ground-breaking factory prototype concept car was conceived by Gordon and designed with help from Claude Hill, the premier chassis designer, just before the War. It was protected by a series of patents related *inter alia* to its use of a strong, easily modified tubular structure to which external body panels could easily be attached and varied cheaply. With its first outing in the Spring of 1940, the press reports were favourable. "Autocar" described it as "the suggestion of the car of the future" recognising it as a complete break with conventional practice, far ahead of anything in production at the time. Its handling far out-stripped any production car of the time and it was almost silent running, all underpinned by simpler production methods. “The Motor” described it as follows; “Technically it demonstrates beyond doubt the possibility of building a car which will provide the performance and the comfort in riding which have previously been associated with cars of twice this nominal power and weight.” The car was used daily, if petrol rationing allowed, both as the factory testbed by Gordon and as his personal transport the family – his son Ian Sutherland always remembered sitting in the back on his way to school. A new four cylinder pushrod engine, capable of even easier manufacture, was under

Dr Tom Rollason (four times winner of the Sir AM Sutherland Trophy) presents Gordon Sutherland with a new bonnet motif, of his own design, for the refurbished Atom. By Permission of Dr Tom Rollason.

development by Gordon in 1944 but never came to fruition. The "Atom" was the testbed for what became the future DB1 Engine. Wartime meant that the original "Atom" never went into production, but a test drive showing the design and performance potential of the car was the deciding factor in David Brown purchasing the company after the War.

The "Atom" was brought back to life in 1996 after a ten-year restoration by Dr Tom Rollason, an Aston Martin Owners Club enthusiast, who fortuitously had won the Sir Arthur Sutherland Memorial Trophy for Best Kept Pre-War Car of the Year four times. It was unveiled at a celebratory event at the factory at Newport Pagnell attended by Helen and Gordon Sutherland who enjoyed a ride in it through the town. Dr Rollason presented the Sutherlands with his newly designed car bonnet motif for the "Atom". Still highly prized, the car recently passed from Dr Rollason in 2016 into the private custodianship of a discerning Northern Irish collector.

During the War the works had switched over to wartime production and suffered bomb damage. However, Sir Arthur anticipated little post-War demand for elite cars and though he had transformed the company, he began to perceive it as a real drain on BJ Sutherland & Co. As ever, Sir Arthur knew when to sell and in a letter to his son dated 12 September 1944, he offered to sell the company to Gordon for the nominal amount of £5! By this time Sir Arthur was no longer a Director and technically he had no controlling interest. However, he owned BJ Sutherland & Co., the controlling company, and probably still held shares. Sir Arthur offered Gordon further advice in terms of his having a sound legal case to be released from the lease at Feltham in that the premises had been badly bombed. He also suggested that Gordon sell the machine tools which were by then in short supply elsewhere. Gordon accepted the £5 offer one month later but did not initially take all of his father's advice until 1947. For the first time Gordon had absolute control though without his father's financial resources to produce new models. But Gordon also possessed the family eye for good business, and he advertised the company for sale in "The Times" and sold it in 1947 for £20,500 to David Brown Ltd., the Huddersfield precision engineering manufacturers of machine tools, gearboxes, hydraulic systems and tractors. This brought "the Sutherland era" to an end. Perhaps Aston Martin was but an unusual glamorous indulgence on the behalf of Sir Arthur for his son even though it meant that BJ Sutherland & Co. took a considerable hit on its holdings.

In his novel "From Russia with Love" Ian Fleming had James Bond driving a DB MK III which led to the famous David Brown DB5 model featuring in the James Bond film "Goldfinger." The timeline is, of course, out of kilter but Sir Arthur would surely have approved of the concept of the spy with the Irish name from Scotland whose car would never have been an Aston Martin if he had not bailed them out. The Sir Arthur Sutherland Memorial Trophy for Best Kept Pre-War Car of the Year is still contested annually by the Owners Club. Sir Arthur never drove them; his cars of choice were the luxury Belgian Minerva and two Rolls Royces with the vanity number plates AMS 1, 2 and 3 (long since transferred to Lord – **A**lan **M**ichael - **S**ugar).

The Aston Martin Atom.

Political honours

Newcastle upon Tyne

He followed his father into civic life and was elected a Newcastle City Councillor for the Jesmond Ward in 1910 and became a magistrate in 1916. He then served as Deputy Lord Mayor and Sheriff 1916-17, some twenty-five years after his father had held the post of Sheriff; he was recommended for the position on the grounds of his being "young, able, intelligent, wealthy and good-looking"! He was Lord Mayor 1918-1919 and could do little wrong in Armistice Year. He remained an alderman until his death at which point, he was the "father of the Council" and still attended Council meetings albeit sitting quietly at the back of the Council Chamber.

The Right Honourable Arthur Munro Sutherland JP Lord Mayor of Newcastle upon Tyne 1918-1919.

By John Hodgson Campbell.

He received the Honorary Freedom of the City of Newcastle in 1936. The ceremony was held at the City Hall and he invited all the pupils of the Royal

Grammar School. He published his acceptance speech, "Industry – Thrift – Ambition", and presented it to the School. Said Honorary Freedom (as opposed to his being a Freeman of a Newcastle Guild which includes obligations to defend the City and is hereditary) is the highest honour that the City can bestow and is only granted occasionally but gives no entitlements to the holder and is not hereditary.

THE CITY'S GIFT TO SIR ARTHUR.

The gift to Sir Arthur on his being presented with the Freedom of the City of Newcastle.
By permission of the Sutherland family.

Knighthood & Baronetcy

In November 1918, having just been elected Lord Mayor, he had conferred the Honorary Freedom of the City on PM Lloyd George (who stayed at Thurso House where they all sang Welsh hymns together with Arthur playing the organ in the games room!), Earl Haig and Earl Beatty (the latter two being ecstatic regarding Newcastle's contribution to the War effort not least in coal production and ship repairs). Arthur was soon appointed Knight Commander of the Order of the British Empire (KBE) in the 1920 Civilian War Honours for his services as Lord Mayor, an honour bestowed on various Lord Mayors of England.

The Sutherland Baronetcy of Dunstanburgh Castle in Embleton in the County of Northumberland was created the following year in the 1921 Birthday Honours. The honour of the baronetcy, that of hereditary knights, is a special honour awarded by the British Crown without the privileges of a peerage but ranking higher than a knight. Sir Arthur paid £275 [£9,400.00] to the College of Arms for his Designation in Letters Patent of Baronetcy. The Sutherland family recollect that the baronetcy was awarded for Sir Arthur's role in the "Baltic Escape" but there is no mention of this in the official citations.

Northumberland

In 1925, he was commissioned as Deputy Lieutenant of Northumberland, which post he held until his death. In 1941, he was appointed as one of the three Sheriffs for Northumberland and from 1943 he held the position of High Sheriff of Northumberland involving accompanying the judges to the Assizes, giving notice of the holding of the Assizes and empanelling juries.

City of London

Arms of the City of London.

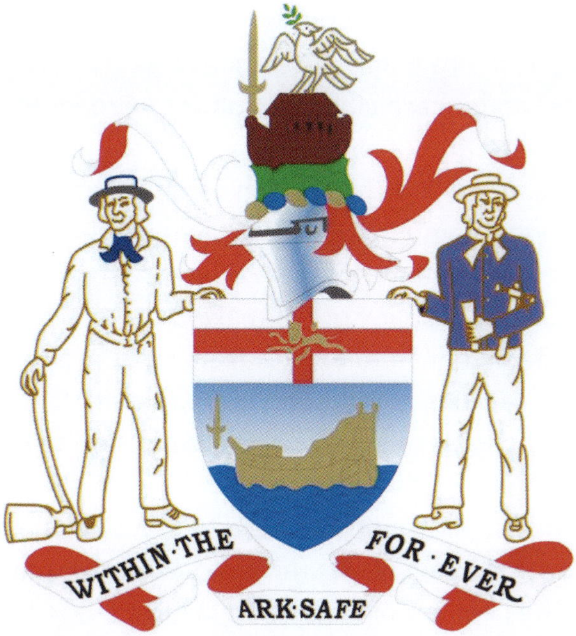

Arms of the Worshipful Company of Shipwrights.
Reproduced with the kind permission of the Company.

Arms of the Worshipful Company of Feltmakers.
Reproduced with the kind permission of the Company.

Whereas there is a difference between being given the Honorary Freedom of a City as opposed to being a Freeman of a City, it is surprising to note that eighteen years *prior* to his receiving the Honorary Freedom of the City of Newcastle, he had been admitted, from 1918, as a Freeman of the City of London. By necessity he had first to have been admitted as a Liveryman (member) to City of London Companies. He was admitted to two Companies; the Worshipful Companies of Shipwrights by Redemption (purchase) and Election and the Worshipful Company of Feltmakers on 10 May 1918. There were only twenty-one Liverymen in the Feltmakers in 1918 and so there is significance in one of them being a non-Londoner. Joining the Feltmakers may seem an unusual choice but was reflective of the use of felt as insulation and packing for ships' boilers, pipes, insulation, gaskets and washers all prior to the widespread use of asbestos. Election to the Liveries was for life though he took the highly unusual step of resigning from the Feltmakers in April 1938. Sir Arthur did not progress from the position of Liveryman in either case and his involvement with the Companies was minimal. Nevertheless, at that time the maximum number of Liveryman per Company then permitted by the Court of Aldermen was three hundred and so these were appreciable honours for a Tynesider. His sponsors are unknown as the relevant Minute Books were destroyed in the Blitz. These honours perhaps reflected the role of Newcastle in sending coal to London for lighting, heating, coke making, power production and rail transport.

Order of St John of Jerusalem

Badge of a Knight of the Order of St John of Jerusalem.
By permission of the Museum of the Order of St John, London.

Priory Church of St John, Clerkenwell, London EC1. Image accessed from museumstjohn.org.uk

The King sanctioned his appointment as a Knight of this Order in 1922 prior to which he had been invested successively as Member, Officer and Commander in the Priory Church of St John in Clerkenwell, London EC1. His investiture as a Knight when he declared his allegiance to the Order would have been in a Royal Palace. The qualities required of a Knight are "Chivalry and Care, Valour and Faith." Investiture into this Victorian Order of Chivalry is by invitation only and is dependent on philanthropic concern in medical and health matters. This honour reflects Sir Arthur's concern with the blind and war disabled. He had supplied ambulances for the Western Front and founded the Sutherland Hostel for Wounded Soldiers in Newcastle. He was Chairman of the Tyne and Blyth Benevolent Fund which assisted Old Quaysiders and was also involved in the funding of three war memorials in Newcastle.

Durham University; Doctor of Civil Law

In 1932 the University of Durham conferred this Honorary Degree upon him for his philanthropic work in the North East and for his funding of the Newcastle Medical and Dental Schools in 1930 and 1931 which were then (and until 1963) part of Durham University. It was conferred in the Chapter House of Durham Cathedral.

"Arthur Munro Sutherland in the robes of Doctor of Civil Law Durham" by Frank O. Salisbury.
By permission of the Estate of Frank O. Salisbury.

Norway; Order of St. Olaf

Badge of a Commander of the Order of St Olaf.

In 1936 the King of Norway, Haakon VII, conferred upon Sir Arthur the honour of Commander of the Order of St Olaf [sometimes Olav] (First Class with Star) "for remarkable accomplishments for the country and humanity" as confirmed by Haakon and the Norwegian government. The ceremony took place at Thurso House and the honour was presented by Paul Wisness, the Norwegian Consul for Trade as appointed to the North East of England. Specifically, it was in recognition of his fostering mutual trade of coal, timber, iron and tourism for fifty years including the building of the special quay facilities at the Malmo Wharf for Scandinavian trade under his auspices as Chairman of the Tyne Improvement Commission. In other words, it was granted for the reciprocal Scandinavian trade that Sir Arthur had fostered and for the construction of the Tyne Commission Quay to foster Anglo-Norwegian passenger trade.

Homes

"Eastcliff"

Arthur was born in Thurso House, lived most of this life there and would die there. However, in 1893 he had married Fanny Linda Haggie and they set up home at "Eastcliff", 80 Osborne Road, Jesmond. They would move back into Thurso House in 1915 subsequent to Arthur purchasing it from his sisters to whom it had been left in their father's Will but who had not had the financial wherewithal to maintain it.

38-40 Sandhill, Quayside

Arthur and Benjamin John the younger, his eldest brother, inherited numbers 32, 33, 34, 38, 39 and 40 Sandhill (shops, warehouses, offices and a 'dwelling house') from their father in 1901. Arthur became sole owner in 1908 following the death of his brother. The original grain warehouses are now the Quilted Camel Retro Bar. The original offices at 38-40 (two doors from Bessie Surtees House)

BJ Sutherland & Co. boardroom, 38-40 Sandhill. By permission of the Northumberland Press Ltd.

became his own business headquarters and, in many ways, were his *real* home. These were the original merchants' houses of the sixteenth and seventeenth centuries with their timber-framed frontages and arrays of windows showing considerable wealth as glass (a Tyneside speciality product) was then expensive. Photographs of the interior of his office show it as being almost identical to that of Bessie Surtees House now. A portrait of his father hung on the wall. This was the ideal place to impress business associates and rivals. The various floors were his offices, personal residence, function rooms and servants' quarters on the top floor. He would duplicate many of the details of this residence in both the fireplace in the reception room at Thurso House and at Sutherland House, his London Headquarters of the Free Trade Wharf.

The 'Change and the Union Club

Whereas these organisations were not his home as such, Sir Arthur conceded that he frequented them on virtually every working day in Newcastle. The two were in sharp contrast to each other, the former exuding hustle and bustle and the latter the calm of an elitist gentleman's club. The clientele differed but the object was the same; to cut deals and to keep abreast of commerce nationally and world-wide for the benefit of the North East.

Commercial Exchange, Newcastle Quayside. 1946
By permission of Newcastle City Library.

His personal offices were strategically placed directly opposite the Guildhall, the ground floor of which was then the Chamber of Commerce and the Commercial Exchange known to local traders as The 'Change. Traders used the phrase "on 'Change", meaning at (or going to) the Exchange and Sir Arthur wrote that "for almost sixty years I have been going almost daily "on 'Change". A former Newcastle City Guide was a runner there in the 1940s and 1950s and remembers Sir Arthur and other shipping brokers and coal producers standing in set positions on the floor. A runner's job was to run around with pieces of paper, matching loads of coal with ships. At about 11 am, a bell was rung, and no-one else was allowed in. Many of the traders wore top hats and kept their contracts in them, held in place by elastic straps.

Almost inevitably he would become its President in 1919. Its location in the Sandhill enabled him to take the pulse of industry on Tyneside and have his finger on the button for his transactions and deals. The 'Change was packed with colliery owners, shipowners, ship-brokers and coal exporters and Sir Arthur just had to cross the road to be there. (One hopes that in those transactions that he was aware that the shadow of John Wesley was cast over him whenever he entered, given that Wesley had preached often from a niche in the previous Guildhall on that site.)

The Union Club (now the Union Rooms Bar, Kitchen and Pizzeria adjacent to the George Stephenson Plaza close to the Central Station) was only ten minutes away but was a world apart. It was an elite gentleman's club; "Membership is very exclusively for local, wealthy and like-minded businessmen." In construction it appears as a French chateau and within are fine staircases, stained glass (featuring poets, portraitists and musicians), smoking rooms, a billiard room, dining rooms, meeting rooms and bedrooms for overnight accommodation. Membership was strictly controlled and was by introduction of two members and a vote of the entire membership; Newcastle Members of Parliament were members. Precise preparation times were given next to each item on the menu to enable sitters to plan their day. Heavy industry dominated The Change, but the Union Club was ideally located to serve the banks and insurance companies that then lined Mosley Street and Collingwood Street. Sir Arthur could glide from one world to another in the middle of the day. (Sir Arthur was a club man as he also held membership at the Bath and Reform in London.)

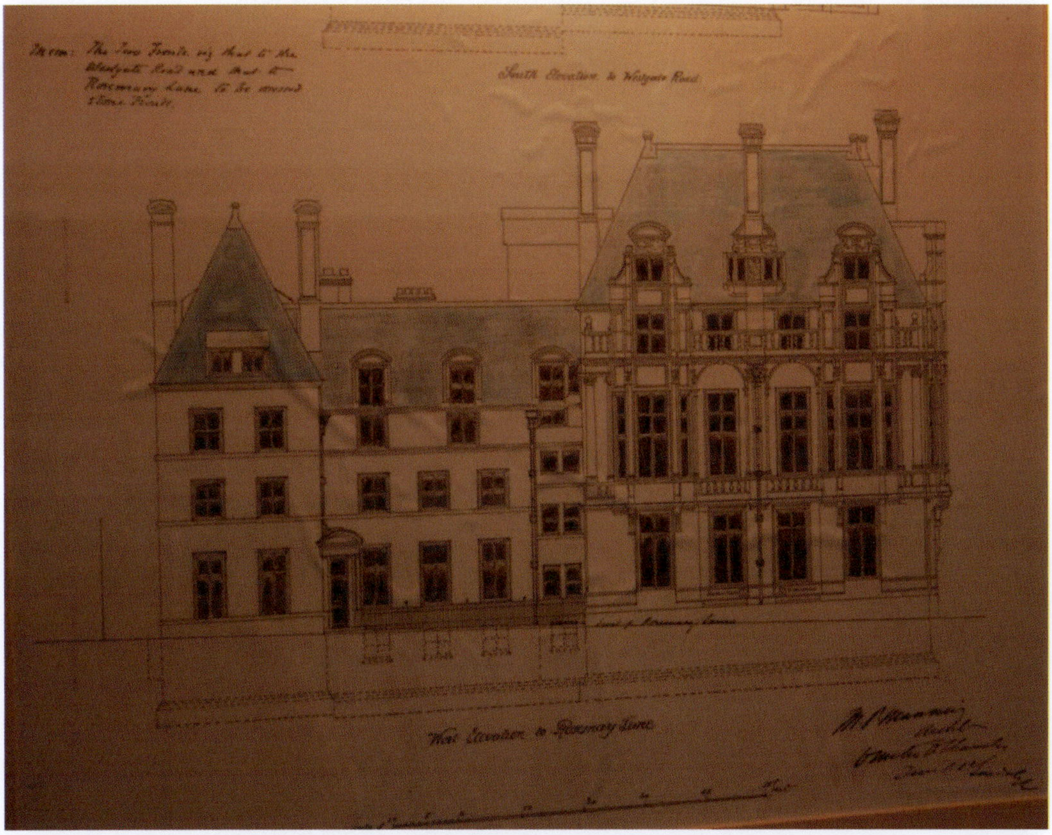

West elevation plan of the former Union Club of Newcastle. "A French chateau in the middle of Newcastle." By permission of the Union Rooms Bar. Photograph by Nigel McMurray.

Thurso House (and Kelso House)

This house, in Fernwood Road, Jesmond is now the Mansion House. The name is significant in that the Sutherlands' origins were in the then destitute town of Thurso, though Thurso has long since forgotten them.

It is often mistakenly assumed that Thurso House was constructed by Arthur as his town house, which he later bequeathed to Newcastle after his death. This is only partially true. The original structure was in fact two adjoining but separate homes under the one title of Keslo House, built in 1879 as a high class semi-detached villa. [See page 13 for further details.] Both homes were purchased by Arthur's father, Benjamin John, in 1879 and the one in which his family lived was given the additional name of Thurso House. Arthur lived in Thurso House as a child and until his marriage. His father had left it to Arthur's sisters in his Will of 1901, but they did not have the wherewithal to maintain it and so Arthur purchased it from them in 1915. He then set about expanding it and improving it on a grand scale including the construction of the present large stone entrance porch with its balconies. The stained glass lamp shades in the entrance porch feature the arms of Newcastle and the walls are 'decorated' with stags' heads from his country retreat in Hethpool. Properties adjacent to it and behind it were purchased and demolished and replaced with gardens and five garages for his cars and those of important guests. Thurso and Kelso houses were joined and became one house named simply Thurso House. The vast

Civic Banquet for the Consular Corps. By permission of the Newcastle Mansion House.

garden behind Thurso House remains though the grand garden parties of yesteryear are a thing of the past.

The Newcastle Consular Corps., which is still extant, is composed of overseas diplomats, importers and exporters based in the North East. Over thirty nations have been represented since its foundation in 1909 and BJ Sutherland Ltd traded with many of them. The Corps. met primarily in the former Mansion House in Saville Row (now Saville Place) and occasionally in Thurso House. Sir

Arthur displayed a photograph of the Corps. in Thurso House to show his intimate links with members of the Corps. Arthur's favourite Doyennes (Presidents) were the Dane, Louis Zollner, who gave the introductory speech for Sir Arthur's being granted the Honorary Freedom of Newcastle. And the Norwegian, Paul Wisness (a Knight of the Order of St Olaf), who recommended to the king of Norway that Sir Arthur be inducted into the Order. In due course, it would be Wisness who inducted Sir Arthur at a ceremony in Thurso House.

Much of the bedroom furniture, which is still in place, was brought from his ships when they were scrapped. A print of the John Wilson Hepple painting of "John Wesley preaching at the Sandhill" – directly opposite what would become Arthur's offices in Sandhill – still hangs in one of those bedrooms. There are photographs of him at ceremonial events hanging on the walls of the upper floors. There are small stained glass panels of heraldic badges and mottos in the upper section of some windows relating to his grandparents' home county of Caithness, his shipping companies and the towns – such as Leith – where his ships were constructed. As a child he had been fascinated by clocks and watches and this hobby developed into a significant private clock collection of which eight long case and two mantle clocks are still on display. It seems that Sir Arthur had always envisaged Thurso House as becoming the new Mansion House in that he collected paintings of the original Quayside Mansion House of 1691-1836, which are still displayed in the Mansion House dining room. They were grand days at Thurso House with his love of gracious living being reflected in there being uniformed staff to serve dinner. Two of the paintings are by the Methodist artist Henry Purlee "Smuggler" Parker (1795-1873) most famous for his depiction of the "Epworth Fire" in which John Wesley was almost killed as a child. (He was nicknamed "Smuggler" as he had inadvertently produced many portraits of smugglers while painting genre scenes of the North East coast!) Parker has two large canvases at Thurso House; "Fancy dress ball in the Mansion House celebrating the Coronation of William IV [1830]" and "Sandhill wine pant on the Coronation of George IV [1821]". Thomas Miles Richardson's "The Sheriffs Procession to meet the judges at the Mansion House" is also in the collection (Sir Arthur was sometime High Sheriff of Northumberland and accompanied the judges to the Assize) as is John Wilson Hepple's "1771 Barge Day". Both Richardson and Hepple were from the North East and Parker

"The Mansion House, Newcastle upon Tyne" taken from an original watercolour painting by artist Alan Reed. www.alanreed.com

had adopted the North East as his home. Perhaps, as an almost vain attempt to offset all the splendour and expense of those events, Sir Arthur also hung a Ralph Hedley painting of "The Post Box" featuring a barefoot boy in the rain.

As expected, Sir Arthur *did* "offer as a gift" Thurso House to Newcastle in 1953 for the purpose of (the City's third successive) Mansion House which role it still fulfils. One short paragraph in his Will finally confirmed this, though the Corporation were to be given just three months to decide if they wanted to accept the gift, after which Thurso House was to be sold and the proceeds given to his family. In death as in life, Sir Arthur needed a snap decision.

Hethpool

Sir Arthur purchased two swathes of land in north Northumberland in the years following the First World War, most of which was let. The estates discussed here stretched from Yeavering near Wooler, going west to the Scottish Border along the College (or College) Valley with the focus on Hethpool. This land was composed of farms, hill farms, cottages, stables, fields, grass lands, quarries, gardens, allotments, boathouses and two hotels. For the second swathe of land see **Other business interests** **Dunstanburgh Castle**.

Hethpool with Sir Arthur at the entrance. By permission of the Northumberland Press Ltd.

The Hethpool Estate was the most important aspect of the College Valley and Hethpool House at Kirknewton became his country residence. Originally constructed in 1687 it was largely ruinous. He purchased it for £16,000 by auction at the County Hotel, Newcastle from the then late Earl Grey in 1919 and re-furbished it in the same year. The work was carried out by George Reavell of Alnwick, one of the North East's most accomplished domestic architects, on the site of the ruins of the original 1687 house. Reavell had to start virtually from scratch and so Sir Arthur arranged for the construction of a narrow gauge private railway line from Kirknewton to Hethpool for the transporting of the stone. Art historians consider it to be of significance though in reality it is but a sturdy farmhouse, albeit with an Arts and Crafts garden and adjacent Arts and Crafts workers cottages. The house and all the land within the College Valley Estate (one of five cuttings into the Cheviot Hills) had originally been given by the nation to Admiral Lord Collingwood who was perhaps the *real* victor of Trafalgar.

Hethpool workers cottages. Photograph by Nigel McMurray.

The significance of this would not have been lost on Sir Arthur given the legendary status of Cuthbert Collingwood, 1st Baron Collingwood. Surely it was time for *another* North East sea-faring legend to inhabit this house and own this land? Collingwood's title was Baron Collingwood of Caldburne and Hethpool. Certainly, Sir Arthur was fascinated by Collingwood in that a lock of Collingwood's hair was – until its theft – on display at Thurso House. And Collingwood and Sir Arthur were both educated at the Royal Grammar School. Sir Arthur would name one of his ships "Hethpool".

Sir AM Sutherland Cup for the Best Black Headed Sheep. Photograph by Nigel McMurray.

This was, and remains, border farming country and the estate remains principally concerned with sheep rearing. Indeed, just over the border is Yetholm in the Scottish Borders, where the "Sir A. Sutherland Cup for the Best Black Headed Sheep" is still contested for annually at the Yetholm Border Shepherds' Show. He first presented the silver cup, made in Newcastle by Reid & Sons of Newcastle in 1946.

He also constructed four bridges to improve general communications in the Valley one of which is still named the Sutherland Bridge.

The Elsdon Burn was damned to form a lake from which the burn tumbles down over a series of cascades serving the purpose of a fish ladder. However, the main purpose of the dam was to generate hydro-electricity for the estate via a generating station still extant; the lake held five million gallons, the volume being required to operate the turbine. Further embellishment of the house was carried out by the Scot (albeit based in Newcastle) Robert Mauchlen in 1928 with the addition of a distinctive conical roofed tower in the Scots Baronial Style to which is attached Sir Arthur's arms. Mauchlen also constructed the row of four farm cottages for the estate workers in the Arts and Crafts Style, now used as holiday lets. (Sir Arthur had previously engaged Mauchlen to construct The 'Change War Memorial.)

Sir Arthur had an octagonal gazebo for small dinner parties constructed outside the main entrance and overlooking the lake. This features a sailing ship as a wind vane and the door is "protected" by a Scottish wildcat rampant carved into the lintel – the wildcat of the crest of Clan Sutherland. The gardens were occasionally used for garden parties. In the 1920s he entertained his sea captains there in circumstances that today would be called an "away day", showing that he was an enlightened employer.

Four small 1812 cannon from the Newcastle Keep – for firing on ceremonial occasions – are placed strategically at the courtyard entrance though there is no record as to how they got there. The original 1687 date stone was saved and has been incorporated into the House.

However, the Hethpool Estate of 7,845 acres caused Sir Arthur so many problems that he considered disposing of it on various occasions. Problems arose from constant outbreaks of lamb disease between the Wars and the annoyance of his tenant farmers that he was doing little to control the disease. Rent disputes were time consuming with one such dispute by the tenant of Elsdon Burn Farm continuing for a year. Even the lake, with its idyllic boat house, enchanting seven step fish ladder and summer house was deemed to be a reservoir because of its size. Inspections under the then Reservoir Act resulted in it having to be reduced in size as it had expanded beyond its original agreed contours. New spillways were also required. Arguments over the potential listing of Neolithic artefacts and intrusions by the Ordnance Survey were further distractions. Sir Arthur then gave the Hethpool Estate as a wedding gift to his daughter Linda Kathleen in 1928 when she married Henry Armstrong, one of his Directors of his Tyne Tees Shipping Co. The Armstrongs could not cope with it either and so, eight years later, Sir Arthur took it back! It had now cost him £60,000 over and above the original purchase price. He made an unsuccessful attempt to sell the estate to the Forestry Commission for the planting of trees in the College Valley but "as usual they have wasted a great deal of my time and in my judgment, they are a broken reed". The house was listed with a London estate agent in Berkley Square, but with markets still recovering from the Depression there was little demand for such country estates. It appears that many of these problems gradually resolved themselves after 1945 with the laying of mains water pipes and his employing skilful estate factors.

But even in 1945 there were financial issues to his detriment in relation to Church of England tithes levied on landowners for the upkeep of its churches. A Tithe Redemption Document of 1945 showed that Sir Arthur had to pay £12,997 [£532,000] to the Tithe Redemption Commission in order to release him from annual tithes of £509 [£21,000].

Hethpool was sold by auction shortly after Sir Arthur's death. The auction was held at the County Hotel where he had first purchased it and the catalogue was almost a replica of the original auction. It is now a bed and breakfast establishment managed by the granddaughter of Sir Arthur's lead factor. The estate is now managed by the College Valley Estates.

Crookham, Northumberland

A further example of his constant philanthropic involvement in whatever situation he found himself was manifest in this small village a few miles from Hethpool. Its historically significant Presbyterian chapel was built there in 1745 for the migrant Scots agricultural workers who criss-crossed the border and had little sympathy for the '45 Jacobite Uprising. Sir Arthur rebuilt and expanded this chapel in 1933. It is now a Chapel of Peace as it is close to the Flodden Battle Site.

Portraits

Sir Arthur's Nonconformity, his Scots origins and his devotion to Tyneside were here manifest in a most unusual way, in that he only sat for artists who had connections with one or more of these three emphases in his life.

"The Right Honourable Arthur Munro Sutherland JP Lord Mayor of Newcastle upon Tyne 1918-1919" was painted in 1918 or 1919 by **John Hodgson Campbell** (1855-1927), the Newcastle-born portrait and landscape painter of Scots ancestry. This is located in the Mansion House. Some of Campbell's other portraits are in the city's Laing Art Gallery. (For this image see **Political Honours**.)

A *photographic portrait* by **Walter Ernest Stoneman** (1876-1978) also celebrated Sir Arthur's year as Lord Mayor. In 1917, Stoneman – a professional photographer and Congregational lay preacher – launched the National Photographic Record of all prominent members of British society in collaboration with the National Portrait Gallery. Sir Arthur was already deemed to be such.

The **Alfred Kingsley Lawrence** (1893-1975) portrait of *Sir Arthur Munro Sutherland* of 1928 depicts a noticeably young and confident Sir Arthur sitting on the gunwale of a ship in a heavy sea, holding a plan of a ship. Lawrence was from Sussex but had studied at the King Edward VII School of Art then within Armstrong College in Newcastle. (The building now hosts the Hatton Gallery.) Some of Lawrence's murals of 1940, depicting the history of Newcastle, are in the city's Laing Art Gallery. He was elected as a Member of the Royal Academy in 1938. Though the circumstantial evidence suggests that this is indeed a portrait of Sir Arthur, the Sutherland family have no knowledge of it. (For this image see **Shipping Companies**.)

Sir Arthur's favourite portrait of himself was that of *"Knight Commander with Star of the Order of St Olaf of Norway"* painted in 1931 by the Dundee-born artist **Howard Somerville** (1873-1952). [His full name was Howard Somerville Anderson though he never wrote Anderson when signing his works.] He was "one of the ablest portrait painters that Scotland produced" according to Dundee's McManus Art Gallery. Somerville made his reputation in London as a cartoonist for "Punch" and was a member of the Arts Club, the Chelsea Art Club and the Royal Society of Portrait Painters. He exhibited frequently at the Royal Academy though he was never an Academician (Member). He was a portraitist to the elite and fashionable in the UK, USA and Paris; he was awarded the Silver Medal (1928) and Gold Medal (1939) at the Paris Salon. His original portrait of Sir Arthur hangs in the Royal Grammar School hall, alongside other former pupils Lords Collingwood, Eldon and Stowell. The first copy, painted in 1932, was for the boardroom of Sutherland House at the Free Trade Wharf; its current location, if any, is unknown. The second copy, painted in 1936, is still located in a bedroom in the Mansion House, symbolising Newcastle's strong links with Norway, many of those links being first fostered by Sir Arthur. That bedroom is the one allocated to Norwegian Royalty on their now rare visits to Newcastle; a more frequent and annual user is the Mayor of Bergen when gifting a Christmas tree to the city.

Knight Commander with Star of the Order of St Olaf. By Howard Somerville. 1931.

The acclaimed Methodist artist, **Frank O. Salisbury** (1874-1962), was, to all intents and purposes, the Court Painter to the House of Windsor and he painted Sir Arthur thrice. In 1940 Sir Arthur was one of seventeen members of the Chamber of Shipping who featured in the large canvas *"King George VI visits the Chamber of Shipping"* of which Sir Arthur had been President in 1930. Other Chamber members from the North East who are depicted are Robert Stanley Dalgleish, Viscount Runciman of Doxford and Philip Runciman (all of whom lived in, or adjacent to, Fernwood Road in Jesmond). Until recently the picture hung in the Council Room of the Chamber of Shipping, but due to its premises being down sized, it has been returned to the Haldin family, Sir Philip Haldin being the original sponsor of it. In 1945 Salisbury painted the first of two portraits, *"Arthur Munro Sutherland"*, which is in the dining room of the Mansion House. This painting shows that Sir Arthur did not get everything his own way in that the caption, "Deputy Chairman 1930-1935. Chairman 1935-1945", clearly shows that it was intended for the Royal Grammar School where he expected it to be displayed in its hall. However, the School would only accept a much smaller portrait for the hall and so he had to hang it in his dining room at home instead. It hung first in the Tyne Improvement Commission. (For this image see the **Front Cover**.) Then, in 1948 Salisbury painted *"Arthur Munro Sutherland. Painted in the robes of Doctor of Civil Law Durham"*. Though the DCL was granted in 1932 this portrait was not painted until much later to coincide with the opening of the new home of the Newcastle Dental School. It was hung briefly in the Royal Grammar School Hall and is currently housed within the Newcastle Medical School's Collection of Art. (For this image see **Political Honours**.)

Methodism

Orkney and Caithness

Methodism – "those who live by the method of the Bible" – was founded by the Revs. John and Charles Wesley and was the dominant force in the eighteenth-century Evangelical Revival. It was, and remains, a mainstream Protestant denomination. However, the new denomination splintered into various sub-groups in the decades after John Wesley's death in 1791 and it was not until 1932 that they united into the Methodist Church of Great Britain and Ireland.

The largest of those Methodist groups in the nineteenth century was the Wesleyan Methodist Church which tended to be conservative and right wing in its secular politics. The second largest group constituted the Primitive Methodist Church which was founded in 1810 having returned to the "primitive Methodism" of Wesley, particularly regarding preaching in the open air. It would have a greater affiliation with the poor and then with the trade union movement. Sir Arthur was in the unusual position of having a Wesleyan father and a Primitive mother and so fully understood the ethos and emphases of each denomination. He had a greater respect and understanding of the Primitives than would be expected of most Wesleyans. He had no significant contact with the other offshoots from Wesleyan Methodism.

Methodism grew at a phenomenal rate in England and Wales during and after Wesley's lifetime and now has over eighty million members worldwide. Progress in Scotland in the nineteenth century was less secure in that the Kirk was not going through the period of spiritual torpor then being experienced by the Church of England. No revival was required or wanted and certainly not from English clergy. John Wesley did not help himself by making remarks such as "I always found it difficult to tell the Scots anything that they did not already know" and "The Scots know everything but feel nothing except the pews on which they are sitting". Nevertheless, much progress was made but could not be sustained due to logistical problems in allocating preachers to the far north of Scotland. In Benjamin Sutherland's early life in Orkney there were Methodist causes in four towns including Kirkwall, and on mainland Caithness there were six causes including Wick and Thurso. He and his immediate relatives were residents of Orkney and Thurso and his wife was born in Caithness. It is inconceivable that they were unaware of those destitute Methodist preachers roaming the land and preaching in the open air. This could have prepared him for its growing dominance in Newcastle when he disembarked there in 1825.

The Haggie family of industrialists were always supportive friends of the Sutherlands from the time of their arrival in Newcastle. They always claimed to be fellow Orcadians and could certainly trace their ancestry back to Robert Sinclair (1725-1814) a master mariner of Kirkwall who emigrated to Newcastle in 1814. The Haggies came under the spiritual spell of the Methodists in Newcastle and became members of the Nelson Street Chapel, financial supporters of the Clayton Road Methodist chapel and were involved in chapel construction as at

Ballast Hills on the Ouseburn. Latterly they sent some of their sons to the Leys Methodist School in Cambridge and retain Methodist links to this day.

The marriage of Benjamin John – a rich industrialist and local political power broker - in 1859 in the Brunswick Wesleyan Methodist Chapel to the daughter of a Primitive Methodist "Ranter" preacher from Gateshead, was an unusual match but ensured the place of Methodism in Sutherland family history.

Clayton Road Wesleyan Methodist Chapel, Newcastle

Sir Arthur stated, "I had the advantage of a sound Nonconformist upbringing." He was a prominent life-long member of the Clayton Road Wesleyan Methodist Chapel in Jesmond at the junction with Eslington Terrace which is now Pilgrims Court Methodist Home [for the Aged]. The original 750-seater neo-Gothic building was completed in 1883 as an overflow from Brunswick chapel and many famous Newcastle Methodist names were its trustees and guarantors - JJ Fenwick, three members of the Bainbridge family and WD Stephens. In due course the Runciman Family would also become members of that chapel.

Bishop Matthew Simpson. From "The Illustrated History of Methodism" by WH Daniels. 1880.

The man chosen to lay the foundation stone in 1881 was Rev Dr Matthew Simpson, senior bishop of the United States Methodist Episcopal Church. He was dedicated at birth to be a Methodist preacher and became a professor, preacher, abolitionist, and confidante of President Lincoln. He preached the funeral sermon at Lincoln's funeral in May 1865 at the Springfield, Illinois, Methodist Church. In 1881 Simpson was in London where he preached the opening sermon at the first Methodist Ecumenical Congress at

Wesley's Chapel at which there were delegates from throughout the world. One of those delegates, Emerson Bainbridge, was from Newcastle.

Simpson preached throughout Britain during his visit including travelling to Newcastle where he met up with Bainbridge and laid the foundation stones at Clayton Road. Ten stones were laid one of which bore Simpson's name and two of which bore the name of Sutherland; Arthur's mother and Jane, her twenty year old daughter. (It is to Benjamin John's credit that he did not put his own name on one of those stones.) One suspects that Arthur and his sister would never forget the day that Simpson came to town and Arthur would have a lifelong commitment to Methodism and Clayton Road.

Bishop Matthew Simpson preaching at the then London HQ of the YMCA in The Strand prior to his visit to Newcastle. From "The life of Bishop Matthew Simpson" by GR Crooks. 1890.

"Clayton Road Wesleyan Church". Pre-1932.
'A Denham Brash Post Card, Heaton, Newcastle.'

Arthur was married at Clayton Road Methodist Chapel in 1893 to Fanny Linda Haggie, daughter of his former boss R. Hood Haggie. The Chapel would benefit much from Sir Arthur's generosity and he stated in the 1933 "Who's Who in Methodism"; "Generous donor to Methodist Churches, including organ at Jesmond." The organ was gifted in 1923 and was constructed by "the celebrated organ builder" JJ Binns of Leeds; it was played at its first morning worship service by WW Ellis the organist at Newcastle Cathedral. In 1929, he provided the

Sutherland Hall that augmented the original church halls and he also paid to repair the damage to the church by air raids in the Second World War. He also gave financial aid to the nearby Methodist International House at 64-68 Osborne Road which was linked with Clayton Road; it is now a hotel.

Sir Arthur was a teetotaller – "all my life I have been practically teetotal" - and a non-smoker. He deemed teetotalism especially important, particularly in business, as the heavy drinking of the day could leave brains befuddled in an environment where none, but the alert profited as prices could fluctuate hour by hour on the floor of The 'Change. And no one was more alert than Sir Arthur at The 'Change.

When he received the Honorary Freedom of Newcastle in 1936 it was stated that "he has made generous contributions to Methodist chapels in Northumberland and Durham". His obituaries make the point that he "always liked singing the rousing Nonconformist hymns." He stated in his autobiography;

> "At that time one of my pastimes was playing the organ as I had done earlier at local missions. I always enjoyed the old-fashioned rousing hymns, which I'm sure did much to stimulate the fervour and religious devotion characteristic of those days. Going to chapel was more than just a duty to most of us. You left with new courage and hope. They were grand days, those chapel days of the last century. Religious and communal life is the poorer for their passing."

Charles Wesley statue, Bristol

Bristol, along with London and Newcastle, were the Wesleys' three national centres. Bristol was chosen as it was the centre of the English slave trade to which the Wesleys were adamantly opposed. A bronze equestrian statue of John Wesley, by Arthur George Walker, was unveiled in the courtyard of The New Room, Bristol – the first Methodist chapel – in 1932. But what of Charles Wesley, "the sweet singer of Methodism"? In 1938 Sir Arthur corrected this omission by commissioning and paying for a bronze statue of him on a limestone base which still stands in the rear courtyard.

Statue of Charles Wesley. Photograph by Neil Jinkerson.

The sculptor was Frederick Brook Hitch. Cut into the front of the pedestal is the statement; "Charles Wesley 1707-1788. 'O let me commend my Saviour to you.'" The reasons were that Sir Arthur was a musician in his own right and the Rev Dr T Ferrier Hulme, who was primarily responsible for the refurbishment of the New Room after securing it for Methodism in 1929, had been the minister in Jesmond when Arthur was in the Sunday School. The statue was unveiled by Florence Sutherland, one of Sir Arthur's sisters, in the presence of the Lord Mayor and Lady Mayoress of Bristol. The unveiling was front page news in the Newcastle Evening Chronicle.

Wesley's Chapel, London

His first wife, Dame Fanny Linda Sutherland, died in 1937; the funeral service was held at Clayton Road Methodist Chapel followed by a private service at Thurso House and internment of her ashes in Jesmond Old Cemetery.

He re-married in 1938. This marriage was to a Danish lady from Copenhagen, Mrs William Robertson Heatley (formerly Bertha Louise Ella Christensen), widow of one of his great friends of thirty years who was a former coal exporter and British Government representative in Denmark. She would have the title of Lady Sutherland. Sir Arthur's London residence was at Whitehall Court off the Embankment and Bertha resided at the Wesley's Chapel campus in City Road, Finsbury. They were married at Wesley's Chapel – "the Mother Church of World Methodism" - on Wednesday morning 19 October 1938. His old friend from Clayton Road and Bristol days, the Rev Dr T. Ferrier Hulme conducted the service accompanied by Rev Percy J. Boyling, the minister of Wesley's Chapel. Poignantly for Sir Arthur, the ceremony was held beneath the stained glass window in the Chapel which commemorates Bishop Matthew Simpson whom he had met as a child at the stone-laying of the Clayton Road Chapel in Newcastle. The bride was given away by Robert Stanley Dalgleish, Sir Arthur's business partner who would be gaoled five years later after the events of the Blyth Docks Bribery Case. Devastatingly for Sir Arthur, his second wife died a short time after the marriage in January 1940.

John Wesley; "The use of money."

John Wesley preaching on the Sandhill by John Wilson Hepple. From "Wesley his own biographer". 1890.

Sir Arthur and Benjamin John, his father, were acutely aware of their Methodist heritage in that they were both linked intimately with the Newcastle Quayside which was a major stamping ground of John Wesley during his fifty preaching and organising visits to Newcastle. Benjamin John's original quayside business premises were at the corner of Broad Chare and the Quayside, barely fifty metres from the old Milk Market where Wesley preached his first sermon on Tyneside. Long hallowed ground for the Methodists, it was commemorated with the erection of an obelisk with three drinking fountains - for people, horses and dogs in - October 1891 (the centenary of Wesley's death) and is still extant in Wesley Square. The Newcastle Courant described the scene attended by the Mayor of Newcastle, Town Clerk and Aldermen alongside Methodist ministers and laymen. Benjamin John had been elected as Sheriff just a few months before and, as the leading Methodist on the Newcastle Council, would have expected to play a leading part in the celebrations. However, due to serious illness this was not possible with the Mayor "apologising for his unavoidable absence".

Latterly, Benjamin John's business premises – which in due course became those of Sir Arthur – moved to the Sandhill opposite the Guildhall. John Wilson Hepple had famously painted Wesley preaching on the steps of the Guildhall (albeit not the Guildhall that Sir Arthur knew but at the same location), the panorama of which was seen clearly from the Sutherlands' business premises. Hundreds of prints were made of the image and Sir Arthur retained one in his bedroom at Thurso House throughout his life and it remains there to this day.

Men like Sir Arthur amassed great wealth which they often channelled back into the community via philanthropy, monuments to their civic pride, monuments to their faith and by developing their passions for collecting. John Wesley disagreed with the accumulation of such vast wealth. Wesley knew that capital was needed, a lesson that he learned again and again when a new chapel was to be built, but unnecessary wealth was always a great concern to him. "Let me never meet a rich Methodist." "The poor are the real Christians." "Earn all you can. Save all you can. Give all you can." "Money never stays with me; it would burn if it did. I throw it out of my hands as soon as possible, lest it should find a way to my heart." Sir Arthur certainly fell well short of Wesley's edicts in these matters in that his personal luxurious lifestyle, like most of his philanthropic contemporaries, remained unaffected whatever the national financial and employment situations.

Philanthropic enterprises

Arthur always kept a "charity account" in his school days with a set percentage being set aside for Barnardo's and local missions. He continued this principle throughout his life with the result that Newcastle benefited on an unprecedented scale.

Royal Grammar School

Royal Grammar School organ. By permission of the Sutherland family.

Sir Arthur was a former pupil albeit a reluctant one. He became a Governor from 1919, Deputy Chairman of the Governors from 1930 -1935 and Chairman from 1935 until his death. He was the School's greatest benefactor and was referred to as "a prince of givers". Much of his philanthropy was offered under Ebenezer R. Thomas' headship from 1922-1948 whose vision was funded by Sir Arthur. His many gifts to the School included a rifle range and armoury in 1915 which is now gone, a James Jepson Binns organ in 1919 in memory of the one hundred and thirty eight Old Novocastrians who died in the First World War, swimming baths (£20,000) in 1931 and only recently replaced, better accommodation for staff in 1936 and a gymnasium (£10,000 but now replaced).

He also met the cost of petitioning for a Grant of Arms [coat of arms] in 1930. This was part of a rebranding process by Thomas and Sutherland to reinvigorate its sense of status and pride. The Letters Patent on vellum remain in the School archives. A new motto *Discendo duces ("*Through learning you will lead") was introduced simultaneously.

He was remembered as being an ever-present figure at Speech Days. The pupils were an integral part of the audience at the City Hall when he was given the Honorary Freedom of the City of Newcastle upon Tyne in 1936. His speech in appreciation of the honour was addressed to the audience in general, but to the pupils in particular. He said that his motto for life was "Industry – Thrift – Ambition" and he would publish the speech and give linen bound copies on handmade paper to the school. He saw the pupils as the future of Newcastle's industrial and financial progress. Rather amusingly, his fame was sometimes celebrated by parodying the words of the School Song with the altering of "Motherland, King and God" to "Sutherland, King and God."

The Old Novocastrians Rugby Football Club was formed in 1898 by former pupils but, having no playing field, it had a nomadic existence. However, by the 1920s it was able to rent some farmland at Benton Lodge in the north of Newcastle, that land being owned by BJ Sutherland & Co. In 1925 Sir Arthur took the seemingly odd decision to give the Club a low interest loan to purchase the land because the Club could not afford the £10,000 purchase price. He was bound by Company Law and he, as Chairman, could not sell an asset of his own company. The Old Novocastrians Association then made him its first President. Then, in 1928, Sir Arthur constructed a new club house, three rugby pitches, a cricket pitch and six tennis courts. As a Methodist, Sir Arthur stipulated that there be no alcohol in that club house though attempts have been made to circumvent this. Benton Lodge was renamed Sutherland Park though it is now known as the Old Novos ground.

Newcastle Medical School

Opening of the Newcastle Medical School 1939. By permission of the Newcastle Mansion House.

The pre-Sutherland Medical School was founded in 1834 and was located in Northumberland Road from 1851.

In 1936 Sir Arthur donated £100,000 to build a new Medical School in King's Road which, on completion in 1937, saw the joining of the Newcastle College of Medicine with Armstrong College to become Kings College within the University of Durham. (In 1963 Kings College would become the University of Newcastle upon Tyne.) It was officially opened by HM George VI in 1939 and was named the New College of Medicine but was soon renamed the King George VI Building by which it is still known. Its red brick façade remains, and it is now (2020) the School of Pharmacy.

Newcastle upon Tyne Dental Hospital and the Sutherland Dental School

Sutherland Building, Northumbria University 2020. Photograph by Nigel McMurray.

In 1931 Sir Arthur gifted £12,600 towards the construction of new premises within the grounds of the Royal Victoria Infirmary [RVI]; this should be seen in the context of a grant of £3,500 from the Dental Board of the UK. The RVI premises were soon found to be too small and so Sir Arthur then donated a further £50,000 in 1948 to move to the original (1851) Medical School building in Northumberland Road. This was opened by the Rt. Hon. Aneurin "Nye" Bevan, the Health Secretary for the National Health Service in the presence of a great gathering including the Vice Chancellor of Durham University, the Lord Mayor, the Sheriff, the Bishop of Newcastle, and the Lord Lieutenant of Northumberland. Sir Arthur was President of the Governors of the Dental Hospital and the building was named the Sutherland Dental School. He gave the School permission to use his arms and motto as its badge on its letter heads. Many Geordies will remember this as the "Dental Hospital." In 1978 came the final move to its current premises in Framlington Place near the RVI. However, the original building that he funded still bears his name as The Sutherland Building, now (2020) the Administrative Headquarters of Northumbria University in Northumberland Road.

Newcastle Society for Blind People

Sir Arthur's elder sisters, Elizabeth (1860-1935) and Florence (1861-1949), were heavily involved with what is now Newcastle Vision Support, which was formerly the Newcastle Society for Blind People. They were generous donors and members of committees and its Council from 1913. Sir Arthur's involvement commenced the year after the death of Elizabeth at which point, he founded the Sutherland Memorial Hall in 1936 in her memory and became a Patron until his death. The Hall, adjacent to the former Presbyterian Church in Jesmond Road, was opened by the Duchess of Northumberland who was a friend of Sir Arthur through his holding high office in Northumberland. The Hall was demolished in 1981. Vision Support now operates from MEA House in Ellison Place.

War Memorials

Great War Memorial, Eldon Square. 1923. From "Official Guide to Newcastle upon Tyne" by Sir AM Oliver. Pre-1939.

Sir Arthur was Honorary Treasurer of the First World War Northumberland War Memorial Committee and organised the "shilling fund" through which donators could each give a shilling subsequent to his initial donation of 20,000 shillings. There was much debate as to the memorial's structure and location, but the result was the monument surmounted by a bronze figure of St George and the dragon in Eldon Square. St George was a popular icon for First World War memorials, and he was the patron saint of the Northumberland Fusiliers. It was unveiled by Earl Haig in 1923 and remains the City and County's principal war memorial. He also donated the total cost of the Newcastle Commercial Exchange War Memorial in the Guildhall which was unveiled by the Duke of Northumberland of which only the plaque has survived attached to a column. He funded other First World War memorials; Clayton

Road Wesleyan Methodist Chapel, Elswick Wesleyan Methodist Chapel and the Royal Grammar School.

Subsequent to the Second World War he constructed the Sutherland Memorial Hall (now Cuddystone Hall) near Hethpool in the Cheviots where both Allied and German aircraft had crashed due to atrocious weather conditions. And invested £78,000 - £1,000 for each year of his life - into Newcastle's Thanksgiving Week, a fund for the regeneration of the city.

War Memorial within the Commercial Exchange. After 1920.
By permission of Newcastle City Library.

Gateshead YMCA

Though his philanthropic gifts were directed in the main to Newcastle there was a notable exception in Gateshead of which Sir Arthur would say, "I have never made a better investment". This was a new YMCA building in High West Street which he named Sutherland Hall after his mother; it was opened in 1947 by Sir Frank Willis, General Secretary of the National Council of the YMCA. Sir Arthur expressed the wish; "I hope that God's blessing will rest on this Centre and its work." The Duke of Northumberland presided over the first anniversary celebrations. An extension was opened the following year by HH Princess Marie Louise. Sir Arthur always had contacts in high places and used them to good effect. This project stemmed from his good friend Thomas Magnay, Methodist local preacher and MP for Gateshead, who ensured that Sir Arthur's magnanimity stretched to the other side of the Tyne with his blunt approach of; "Now listen, you're half a Gateshead man yourself and you're a Methodist and so with a combination like that, don't tell me your Gateshead ventures have ruined you. I want you to buy a Methodist chapel in Gateshead for a YMCA and Youth Centre". This blunt approach worked, and Sir Arthur's response was magnanimous with the construction of that £25,000 YMCA and a Youth Centre in the premises of the former Belle Vue United Methodist Free Church. The main body of the former church became a restaurant which doubled up for Christian outreach meetings, concerts and dances. Upstairs there was a prayer room, library, workshops and a fully equipped gymnasium. It was "one of the finest YMCA centres in the country" and had over one thousand members. The premises thrived and were replaced by larger premises in 1961 at the Bewick Hall in High Street West.

He also established a printing works - the Northumberland Press Co. in South Shore Road - which became a significant employer in the town. It proved highly successful regularly paying dividends from 3% to 6%. It published his autobiography. And he had previously paid the £10,000 cost of repairs to the Gateshead Quay after it fell into the Tyne in 1930.

And finally, his obituary in the Evening Chronicle of 30 March 1953 gives yet another angle on his philanthropy. "He often came privately to the help of those who had suffered in business."

Arms and Motto

His Achievement of Arms (shield, crest and helm[et]) was granted by Warrant from the College of Arms in 1921 and is based on the crest only of Clan Sutherland with its Scottish wildcat *rampant and proper* and the Clan motto *Sans peur* ("without fear"). [Only a clan chief can bear a clan's entire achievement of arms.]

Crest of the Arms of Clan Sutherland.
Copyright of Clan Sutherland Society in Scotland and used with its permission.

Sir Arthur's Arms in stained glass at the Mansion House.
Photograph by Nigel McMurray.

Sir Arthur's Arms are best illustrated in an exquisite stained glass panel in the vestibule of the Mansion House. The two objects placed on the shield are called charges and both are taken from the then arms of Orkney from where the Sutherlands and the Haggies originated. His charge is a sailing ship and the charge of his first wife, Linda Haggie, features a dolphin. The impaling of those arms – the dividing of the shield down the middle - was a common procedure meaning that the husband's and wife's arms appeared equally. There is also the *sinister* (left hand) version of the Red Hand of Ulster on a smaller silver shield denoting his being a baronet; this is seen on the top left hand corner of the shield. The open-vizored steel helm[et] is that of a baronet on top of which is his crest featuring the wildcat of the Clan. There are two mottos; the "Sans peur" of the Clan beneath the shield plus "For God and the Empire" surrounding the shield, this motto being conferred with all knighthoods. He also chose an item of personal significance, the badge of the Order of St Olaf of Norway which hangs beneath the shield. Heraldry is often described as "the shorthand of history" and his arms encapsulate the significant aspects of his life.

Veil Sir Arthur

Sir Arthur and Fanny had four children, and it may seem fair to suggest that he was not a family man, his focus being directed primarily on his commitments in business, philanthropic enterprises and local politics. Certainly, Munro his eldest, left the family circle after a serious dispute with his father over the sale of the Newcastle Chronicle interests. Linda Kathleen, the second child and his only daughter, was initially somewhat of a rebel and recluse apparently "terrifying the neighbours by driving through the city at great speed in a coach with four horses". She maintained a 'flapper' image even when she ostensibly opened Sutherland Park at the Royal Grammar School subsequent to her recent marriage to a director of one of Sir Arthur's companies. Benjamin Ivan, his second son, became a conscientious landowner and farmer of the Dunstanburgh Estate where he constructed high quality accommodation for the workers. Robert Gordon, his youngest son, was an imaginative and gifted engineer who worked closely with, and received financial and financial advice from, his father particularly with regard to Aston Martin. Gordon enjoyed the life with the "jet set" who frequented the emerging world of grand prix racing.

Sir Arthur was very astute in his financial timing. He analysed everything and had all the figures, to the last penny, at his fingertips. He watched the stock market and the Daily Lloyds Lists and noted financial fluctuations and the cargoes coming in and out of the UK. He travelled frequently to the USA to both learn from and monitor the Empire's greatest sea rival. In a letter to his son, Gordon, of 12 September 1944 he reveals his business plan and explains the thinking behind the sale of some of his companies at that point in time. Share prices had recovered very slowly after the 1931 crash only to collapse at the outbreak of the War. Surprisingly, share prices recovered as the War progressed probably on the assumption that the USA would enter the War and victory would be assured. Such was his astuteness that he sold shares in most of his companies in 1944 when their share prices were at their height. Sir Arthur also took the sound advice of his auditor to offset Excess Profits Tax of 100% (this tax was imposed by various Allied governments to prevent companies from profiting from War contracts) by selling Aston Martin, which had made considerable profits due to its manufacturing aircraft components during the War. He also held an extensive portfolio of shares (particularly in railways and private companies running public utilities) and government stocks. Thus, his personal wealth was considerable but so were his philanthropic ventures in Newcastle and his clear intention was that the North East would always be an industrial force to be reckoned with.

During his lifetime, his philanthropic enterprises – which brought him no financial interest - were extended to *organisations* but in death, as per his Will, they were directed to *individuals* (with the notable exception of Thurso House). His Trustees and Executors were Robert Pearson Winter (Chartered Accountant), Thomas Alfred Bertram Forster (Solicitor) and Ernest Ridley (Company Director of BJ Sutherland & Co.). Robert Winter and his father had long associations with BJ Sutherland & Co. as company accountants.

Sir Arthur apologised prior to his death that few gifts would follow given the high rate of death duties and this proved correct in that the nation took £1,583,790 of the £2,013,169 left in his Will. Those who benefited from his Will were his family, his staff at Thurso House and Hethpool including chauffeurs, gardeners and housekeepers. Surprisingly, the major individual beneficiaries, to the tune of £5,000 each, were Thomas and Isabella Wallace his chauffeur and housekeeper at Thurso House. William, one of Sir Arthur's grandsons, recalls that Thurso House was run on "almost military lines" by Thomas, to the extent that he ensured that all of Sir Arthur's clocks kept perfect time and always struck on the hour together! William also recalls the "lavish teas" and "fancy formal dinners" prepared by Isabella. The Wallaces had a remarkable rapport with Sir Arthur and were employed by him for thirty two years; Isabella was a quietly spoken Scot in contrast to Sir Arthur's friends recalling Thomas and Sir Arthur conversing in broad Geordie accents. The Wallaces were guaranteed their residence of Thurso Cottage and its garden (which still adjoins the rear of the Mansion House) rent free for their lifetimes with all utility bills to be paid by Newcastle Council.

His (five) sea captains, chief officers and chief engineers still in service and all members of staff at BJ Sutherland & Co. were beneficiaries. "Loyalty to the firm" was a key factor in the distribution of his wealth in that beneficiaries must have been in post for at least twelve months and were not under notice terminating their employment. Some members of his family were given first refusal to purchase all or any of the eleven farms that constituted his College Valley Estate but none to whom it was offered took up the option. None of that land is currently owned by the family though some family members still live in Northumberland.

There would be a significant tax victory for his Trustees which would have long-term positive consequences. Sir Arthur owned 98,700 shares in BJ Sutherland & Co. on the day of his death up to which point he controlled the company. The shares had to be valued for estate duty purposes by reference to the value of the remaining assets (five ships) and not by reference to the value of his shares. The open market value of the ships at that date was agreed to be £1,150,000. But the ships were sold later at a higher value and so the Inland Revenue Commissioners considered that the estate was liable to more tax. The Trustees argued that the value of the ships at the point of death was accurate and that they, the Trustees, were not liable to pay that extra tax. They lost their case in various courts and finally submitted it to the House of Lords, then the highest Court of Appeal. The case, **Winter (Sutherland's Trustees) v Inland Revenue Commissioners (1961) SVC 227**, was heard in October 1961 before four Law Lords (Lords Reid, Tucker, Hodson and Guest) and Lord Birkett (sometime a Law Lord.) The Court determined in favour of the Trustees and thus the principle was established that the value of assets was to be that at the time of death and said value should neither be dependent on a possible future event, nor to assume that a sale had occurred, nor that it should be dependent on a company's past trading all of which would alter the value. Robert Winter, a Newcastle Chartered Accountant, had argued successfully as confirmed by the Appeal Court's conclusion, with the Lords voting by three votes to two, that the value of assets be that at the date of death. The appeal was confirmed and allowed by Lord Denning as Master of the Rolls which principle is still applies.

Sir Arthur died in his sleep at Thurso House in March 1953 aged eighty-six. He was cremated at the West Road Crematorium and his ashes were interred adjacent to his "Family Burial Place" (which title is cut into the stone) in Jesmond Old Cemetery. A memorial service was held at the Clayton Road Methodist Church on 1 April 1953. The edifice is not ostentatious, but it is very 'busy' with the names of family members. It was originally constructed by his father in memory of Daniel, his father's eldest brother. Those who attend the Cemetery today comment on the contrast between the Established Church section, with its admirable monuments and sculptures, and the Nonconformist section where a measure of constraint was applied. Nevertheless, despite its lack of statuary this is no mean grave and it features the names of his father's entire family, himself, his two wives and some of his family. It was engulfed by foliage for many years which resulted in much damage and makes the deciphering of the names no easy task. There are many Methodists in this Cemetery. Indeed, not far from Sir Arthur is the grave of Tommy Burt, Northumberland Miners' leader who represented the opposite end of the political spectrum. Methodism had taken them in different political directions, yet they share the same graveyard and look upon each other, not in judgement but in blessing.

The Sutherland Family Grave. Photograph by Nigel McMurray.

When the young, poverty-stricken Benjamin Sutherland sailed away from Thurso for Newcastle – with hopes and dreams but without fear - he could never have considered that his little grandson, Arthur, would one day fulfil all those dreams and become The Legend of Newcastle.

Principal Sources and Acknowledgements

Documents

CLAYTON ROAD METHODIST CHURCH, Jesmond, Newcastle upon Tyne. 1881-1983. [1983.] Privately printed at the church.

HUNTER, Inman. "Aston Martin. 1913-1947." Osprey. 1992.

MAINS, Brian & TUCK, Anthony. Editors. "Royal Grammar School. Newcastle upon Tyne. A History of the School and its Community." 1986.

MORGAN, Alan. "Jesmond from mines to mansions." Tyne Bridge Publishing. 2010.

Northern Worthies. 1932. [*Press cuttings*.] Newcastle City Library.

Old Novocastrians Association Magazine. Various entries.

PARRY, David. "A Baltic escape." University of Greenwich (Greenwich Maritime Institute). 2014. Unpublished Masters Dissertation.

ROBINS, Nick. "The Tyne-Tees Steam Shipping Company and its Associates." Bernard McCall. 2014.

SHIPBUILDER & MARINE ENGINE-BUILDER. [Serial.] Various entries.

SHIPBUILDING AND SHIPPING RECORD. [Serial.] Various entries.

SIMMS, Richard. "The Mansion House." n.d.

SUTHERLAND, Arthur Munro, Sir. "Industry – Thrift – Ambition." [No publisher. 1936] **&** Last Will and Testament [1953] **&** "Tynesider." Northumberland Press Ltd. 1947.

SUTHERLAND, Arthur Munro, Sir. to SUTHERLAND, Arthur Munro [son]. [Letter of 1944 on the future of Sir Arthur's companies. Auctioned by Bonhams 2017.]

SWIFT, Wesley. Methodism in Scotland." Epworth Press. 1947.

WEARMOUTH, Robert E. "Methodism and the Trade Unions." 1959.

WINTER-LOTIMER, JB. "Free Trade Wharf, London Docklands." [Unpublished typescript located in Tower Hamlets Archives. 1989.]

People. Ruth Barriskill (ancestry). Peter Butt (Docklands maritime history. Sites and archives in the London Borough of Tower Hamlets). Michael John Dobson (maritime history). Roger Fern (Newcastle City Guide). Bobby Hill (Mansion House). Vivienne Hulme (editing). Alan James (financial interpretations). [Mrs] Eildon Letts (Hethpool). Jane Medcalf and Chris Ward (Royal Grammar School). Rowan Marr (teacher of English). John Peacock (historical comment and editing). Dr Tom Rollason (Aston Martin). Nick Simpson (Sunderland Maritime Heritage). The late Frank Whitehead (legal interpretations and historical comment). Sources that prefer to remain anonymous.

Archives. Gateshead. London Borough of Tower Hamlets. Minneapolis Central Library. National Maritime Museum (Caird Library). Newcastle upon Tyne. Northumberland. Sunderland. Tyne & Wear.

Websites. ancestry.co.uk britishnewspaperarchive.co.uk gracesguide.co.uk

skyscrapercity.com thegazette.co.uk theshipslist.com wrecksite.eu

Cranes on the Tyne. Photograph by Frederick McMurray.